SPECTRAL THEORY AND ASYMPTOTICS OF DIFFERENTIAL EQUATIONS

NORTH-HOLLAND
MATHEMATICS STUDIES 13

Spectral Theory and Asymptotics of Differential Equations

proceedings of the Scheveningen conference on differential equations, The Netherlands, September 3-7, 1973

E. M. DE JAGER
University of Amsterdam

1974

NORTH-HOLLAND PUBLISHING COMPANY - AMSTERDAM • LONDON
AMERICAN ELSEVIER PUBLISHING COMPANY, INC. - NEW YORK

Library of Congress Catalog Card Number: 74 78465

North-Holland ISBN: S 0 7204 2700 2
North-Holland ISBN: 0 7204 2713 4
American Elsevier ISBN: 0 444 10641 3

PUBLISHERS:
NORTH-HOLLAND PUBLISHING COMPANY – AMSTERDAM
NORTH-HOLLAND PUBLISHING COMPANY, LTD. – LONDON

SOLE DISTRIBUTORS FOR THE U.S.A. AND CANADA:
AMERICAN ELSEVIER PUBLISHING COMPANY, INC.
52 VANDERBILT AVENUE, NEW YORK, N.Y. 10017

PRINTED IN THE NETHERLANDS

PREFACE

These proceedings form a record of the lectures delivered at the Conference on Spectral Theory and Asymptotics of Differential Equations held in Scheveningen (the Netherlands) from 3 to 7 September, 1973.
The conference was attended by 40 mathematicians from France, Germany, Sweden, the United Kingdom and the Netherlands.
The number of participants has been limited to 40 in order to give full opportunity for discussions and exchange of ideas. A list of participants with their addresses is to be found on page 207 of these proceedings.
All lectures were given on invitation and the table of contents on page vii gives the titles and the respective speakers.
The Organizing Committee consisted of B.L.J. Braaksma, W. Eckhaus, E.M. de Jager and H. Le mei. The committee thanks all participants and in particular the speakers who made this conference so successful.
The committee is very much indebted to the Minister of Education and Sciences for giving a generous financial support without which this conference could not be held.
The committee thanks also Mr. M.H.J. Westerhoff and Mr. L.E. Leeflang of the Department of Education and Sciences of the Government for their helpful cooperation in financial affairs.
In these proceedings the texts of the lectures have been put in a certain order such that lectures dealing with basically related subjects are brought together (e.g. 1-9 and 15-19).
Since the famous papers by Hermann Weyl in the "Mathematische Annalen" and in the "Nachrichten der kgl. Gesellschaft der Wiss. zu Göttingen" in 1910 on differential equations with singularities, in particular

$$M[y] = -(py')' + qy = \lambda ry, \quad 0 \le x < \infty,$$

there have been published many investigations on this subject, a.o. by Stone, Titchmarsh, Kodaira, Coddington and Naimark. This topic is still a subject of lively interest. The contributions 1-9 give recent developments in this beautiful theory. W.N. Everitt investigates the case where the weight function r occurring in the differential equation may be unbounded or oscillatory; H.D. Niessen and A. Schneider consider so called left-definite systems of differential equations, a generalization of the case

$$|r(x)| \le \rho q(x) \quad \text{on} \quad [0,\infty).$$

J.B. MacLeod investigates conditions in terms of p and q for M to be limitpoint, M.S.P. Eastham gives examples of second and fourth order differential equations with oscillatory coefficient $q(x)$ such that M or its generalization is not of limit-point type and B.D. Sleeman considers a.o. a generalization of the limit-point limit-circle theory to the multiparameter case on which there is also a contribution by F.M. Arscott.

PREFACE

R. Martini deals with differential expressions of the type $\alpha D^2 + \beta D$ with α positive on a bounded open interval I, but zero at the boundary of I. Pleijel's contribution bears upon a positive symmetric ordinary differential operator combined with one of lower order and it is a generalization of his work on limit-point and limit-circle theory.

We have digressed here a little bit on the contributions of these authors because we believe that these contributions together give a kind of "the state of the art" of current research in the Weyl theory on singular differential operators and they may serve as an up to date introduction in this field of mathematical research.
Partial differential operators are dealt with in the papers 10-12; the subjects are degenerate elliptic operators in unbounded domains, scattering theory for wave operators and self adjoint extensions of a Schrödinger operator. Then follow two papers on distribution theory in connection with differential equations (13-14), one on the connection between certain spaces of generalized functions and associated linear operators and the other on quasi analytic solutions of a class of convolution equations. Finally the lectures 15-19 are again a series of lectures on subjects which have a common theme viz. asymptotic approximations of solutions of differential equations. In particular we mention here the paper by W. Eckhaus, who gives an improvement of the well-known Krilov-Bogolioubov Mitropolski method for obtaining an asymptotic approximation for non linear oscillations. The other papers deal with matching principles, singular perturbations for linear and non linear elliptic equations, and weakly non linear oscillations.
Because these proceedings exhibit in coherent sequences of papers modern developments of spectral theory and asymptotics of differential equations, we hope and expect that this book may stimulate research mathematicians and advanced students working in differential equations.
The editor expresses his thanks to Mrs. M. van der Werve, secretary in the department of mathematics of the University of Amsterdam, for her assistance as well during the period of organizing the conference as during the final stage of the preparation of these proceedings.

Amsterdam, February, 1974
E.M. de Jager, Editor.

TABLE OF CONTENTS

1. Åke Pleijel,	A positive symmetric ordinary differential operator combined with one of lower order.	1
2. W.N.Everitt,	Some remarks on a differential expression with an indefinite weight function.	13
3. H.D.Niessen, A.Schneider,	Spectral theory for left-definite singular systems of differential equations I.	29
4. H.D.Niessen, A.Schneider,	Spectral theory for left-definite singular systems of differential equations II. .	45
5. J.B.McLeod,	The limit-point classification of differential expressions.	57
6 M.S.P.Eastham,	Second- and fourth-order differential equations with oscillatory coefficients and not of limit-point type.	69
7. B.D.Sleeman,	Some aspects of multi-parameter spectral theory.	81
8. F.M.Arscott,	Integral-equation formulation of two parameter eigenvalue problems.	95
9. R.Martini,	On differential operators singular at the boundary.	103
10. W.D.Evans,	Degenerate elliptic operators in unbounded domains.	111
11. Kresimir Veselić, Joachim Weidmann,	Scattering theory for a general class of differential operators.	119
12. Ian M.Michael,	The domains of self-adjoint extensions of a Schrödinger operator.	129
13. Magnus Giertz,	Spaces of generalized functions associated with linear operators.	131
14. E.M.de Jager,	On functions holomorfic in tube-domains $\mathbb{R}^n + iC$.	137
15. Wiktor Eckhaus,	On the asymptotic theory of non-linear oscillations.	149
16 J.Mauss,	On first order matching process for singular functions.	163
17. J.Grasman,	The birth of a boundary layer in an elliptic singular perturbation problem.	175

18. A. van Harten, Singular perturbation problems for non-linear elliptic second order equations. 181

19. H.W.Hoogstraten, An asymptotic theory for a class of weakly non-linear oscillations. 197

Author Index 207

Professors F. Stummel and A.H.M. Levelt delivered lectures, whose proceedings do not appear here.

F.Stummel: Pertubation theory for Sobolev spaces and elliptic boundary value problems.

A.H.M. Levelt: Jordan decomposition for a class of singular differential operators.

A positive symmetric ordinary differential operator combined with one of lower order

by

Åke Pleijel

The lecture completes and refines a communication [2] by K. Emanuelsson, Department of Mathematics, Uppsala. Special cases of the problem were treated in [4], [5] and in a so far unpublished joint paper [1] by C. Bennewitz and the author. The method is a generalization of H. Weyl's famous paper [8]. For similar purposes a generalization of the same kind was used in [6], [7]. The written version of the lecture was worked out during a visit to the University of Dundee. For the opportunity to make this visit the author wishes to thank the Science Research Council, United Kingdom.

Two formally symmetric ordinary differential operators S and T are considered of which S has a higher order than T and a positive Dirichlet integral. The operators are given on an arbitrary interval I by sums

$$S = \sum_{j=0}^{m} \sum_{k=0}^{m} D^j a_{jk} D^k, \quad T = \sum_{j=0}^{n} \sum_{k=0}^{n} D^j b_{jk} D^k, \tag{1}$$

where $D = id/dx$ and the complex valued functions a_{jk}, b_{jk} are sufficiently regular on I and enjoy hermitean symmetry. It is assumed that

$$a_{mm}(x) \neq 0 \quad \text{on I.}$$

That the definite order $M = 2m$ of S is greater than the maximal order N of T, in particular means that $n \leq m$. By partial integrations over $J = [\alpha, \beta] \subset I$ a formula

$$\int_{\alpha}^{\beta} Su\,\overline{v} = {}_{\alpha}^{\beta}(u,v)_S + i{}_{\alpha}^{\beta}[\cdots] \tag{2}$$

is obtained by which the Dirichlet integral ${}_{\alpha}^{\beta}(u,v)_S$, or

$$\underset{J}{(u,v)}_S = \int_J \sum_{j=0}^{m} \sum_{k=0}^{m} a_{jk} D^k u \overline{D^j v}$$

is introduced. The Dirichlet integral is a hermitean form determined by representation (1) of S. The dots in (2) indicate out-integrated parts.

For the theory it is necessary to consider the relation $Su = T\mathring{u}$ or more precisely a linear space

$$E(I) = \{U = (u,\mathring{u}) \in C^M(I) \times C^K(I) : Su = T\mathring{u}\}$$

of ordered pairs $U = (u,\mathring{u})$. The sufficient regularity conditions chosen in this definition mean that u and $\mathring{u}$ have continuous derivatives on I of all orders $\leq M$ and of all orders $\leq K = \max(N,m)$ respectively. Then a Green's formula

$$\underset{J}{Q}(U,V) = \underset{J}{[}q_x(U,V)] \tag{3}$$

holds for every compact interval J of I when $U = (u,\mathring{u})$, $V = (v,\mathring{v})$ belong to E(I). In (3)

$$\underset{J}{Q}(U,V) = i^{-1}(\underset{J}{(}\mathring{u},v)_S - \underset{J}{(}u,\mathring{v})_S), \tag{4}$$

while q_x is an out-integrated part containing derivatives of $u,\mathring{u},v,\mathring{v}$. By computing q_x and completing squares it follows that

$$q_x(U,U) = \sum_1^m |\cdot|^2 - \sum_1^m |\cdot|^2$$

so that the signature of the hermitean form q_x, i.e. the pair of its positive and negative inertia indices, satisfies the inequality

$$\operatorname{sig} q_x \leq (m,m).$$

A consequence of (3) is that

$$\operatorname{sig} \underset{J}{Q} \leq (M,M). \tag{5}$$

The solution space $E_\lambda(I)$, where λ is a complex (in general non-real) number, is defined by

$$E_\lambda(I) = \{(u,\lambda u) \in E(I)\}.$$

The elements of this space correspond to solutions of $Su = \lambda Tu$, and $E_\lambda(I)$ has the dimension M. On $E_\lambda(I)$ and $E_{\bar\lambda}(I)$ the form (4) reduces to

$$\underset{J}{Q}(U,V) = c\underset{J}{(u,v)}_S \quad \text{on } E_\lambda(I), \tag{6}$$

$$\underset{J}{Q}(U,V) = -c\underset{J}{(u,v)}_S \quad \text{on } E_{\bar\lambda}(I), \tag{7}$$

where in both cases

$$c = i^{-1}(\lambda - \bar\lambda).$$

A positive character of the Dirichlet integral is essential for the theory. The definition of this positivity is related to a class of compact intervals $J = [\alpha, \beta]$ of which I can be obtained as a monotonic limit. The class only contains sufficiently extended sub-intervals J of I (containing a certain sub-interval J_0). In the sequel the letter J is reserved for intervals in such a class. Dirichlet integrals can be formed with functions having m continuous derivatives on I. It is assumed that for such u the Dirichlet integral

$$\underset{J}{(u,u)}_S \quad \text{is non-negative and non-decreasing} \tag{8}$$

when J increases, and that

$$\underset{J}{(u,u)}_S > 0 \quad \text{on } E_\lambda(I) \tag{9}$$

when u is not identically 0 and λ is non-real (the vanishing on a sub-interval of a regular solution of $Su = \lambda Tu$ implies its vanishing on I).

Because of (9) the reduction formulae (6), (7) show that $c\underset{J}{Q}$ is positive definite on $E_\lambda(I)$ and negative definite on $E_{\bar\lambda}(I)$ provided λ is non-real. In particular $\underset{J}{Q}$ is non-degenerate on these spaces. The $\underset{J}{Q}$-projection $U(J)$ on $E_\lambda(I)$ of an element U in E(I) is determined by

$$U(J) \in E_\lambda(I), \; \underset{J}{Q}(U - U(J), E_\lambda(I)) = 0, \tag{10}$$

where the equality means that $\underset{J}{Q}(U - U(J), V) = 0$ for any V in $E_\lambda(I)$.

The linear subspace $E[I]$ of $E(I)$ is defined by the conditions for $U = (u, \dot{u})$ that

$$(u,u)_{\underset{I}{S}} < +\infty, \quad (\dot{u},\dot{u})_{\underset{I}{S}} < +\infty \tag{11}$$

i.e. that the Dirichlet integrals remain finite when extended over I. From the definition (4) it follows for J tending to I that the form

$$\underset{I}{Q}(U,V) = i^{-1}((u,v)_{\underset{I}{S}} - (u,v)_{\underset{I}{S}}) \tag{12}$$

exists when $U = (u,\dot{u})$, $V = (v,\dot{v})$ belong to $E[I]$. On the similarly restricted solution spaces

$$E_{\lambda}[I] \;=\; \{(u,\lambda u) \in E[I]\},$$

$$E_{\overline{\lambda}}[I] \;=\; \{(u,\overline{\lambda}u) \in E[I]\},$$

reduction formulae of the same type as (6), (7) are valid with the same $c = i^{-1}(\lambda - \overline{\lambda})$, namely

$$\underset{I}{Q}(U,V) = c\,(u,v)_{\underset{I}{S}} \text{ on } E_{\lambda}[I], \tag{13}$$

$$\underset{I}{Q}(U,V) = -c\,(u,v)_{\underset{I}{S}} \text{ on } E_{\overline{\lambda}}[I]. \tag{14}$$

It follows that $c\underset{I}{Q}$ is positive definite on $E_{\lambda}[I]$ and negative definite on $E_{\overline{\lambda}}[I]$. It shall be proved that $c\underset{I}{Q}$ is <u>not</u> positive definite on a linear hull $\{U, E_{\lambda}[I]\}$ in which U in $E[I]$ does not belong to $E_{\lambda}[I]$. In this way $E_{\lambda}[I]$ is <u>maximal</u>, i.e. non-extendable with the indicated property. This statement generalizes a similar one concerning $c\underset{J}{Q}$ on $E_{\lambda}(I)$. While the statement for $c\underset{J}{Q}$ is an easy consequence of (5) the assertion about $c\underset{I}{Q}$ requires a more elaborate proof. A first step is the deduction of the following

<u>THEOREM</u>. <u>To every</u> $U = (u,\dot{u})$ <u>in</u> $E(I)$ <u>for which</u>

$$(\dot{u} - \lambda u, \dot{u} - \lambda u)_{\underset{I}{S}} < +\infty \tag{15}$$

with a non-real λ, there exists a unique $U(I)$ in $E_\lambda(I)$ such that

$$U - U(I) \in E[I], \tag{16}$$

$$Q_I(U - U(I), E_\lambda[I]) = 0. \tag{17}$$

The inequality

$$c\,Q_I(U - U(I), U - U(I)) \leq 0 \tag{18}$$

with $c = i^{-1}(\lambda - \bar{\lambda})$ holds true for every U and its related element $U(I)$.

For an indication of the proof, write $U = (u, \lambda u + f)$ so that $f = \dot{u} - \lambda u$ and $(f,f)_{S\,I} < +\infty$. If $J' \supset J$ a simple computation on account of (4) leads to the formula

$$c\,Q_{J'}(U,U) - c\,Q_J(U,U) = c^2\,(u,u)_{S\,J'-J} - ic\,(f,u)_{S\,J'-J} + ic\,(u,f)_{S\,J'-J}. \tag{19}$$

Because of (8) the Dirichlet integral over $J'-J$ is non-negative. Thus Cauchy-Schwarz' inequality can be used for the last two terms in (19). This gives the inequality

$$c\,Q_{J'}(U,U) + (f,f)_{J'} \geq c\,Q_J(U,U) + (f,f)_J +$$

$$+ \left(|c|\,\|u\|_{J'-J} - \|f\|_{S\,J'-J} \right)^2, \tag{20}$$

where the meaning of $\|\cdot\|_{J'-J}$ is obvious. In (20) the element U can be replaced by $U-V$ with $V = (v, \lambda v) \in E_\lambda(I)$. The difference $U-V$ has the same form as $U = (u, \lambda u + f)$ with the same f. In this way (20) implies that

$$c\,Q_J(U-V, U-V) + (f,f)_{S\,J} \quad \text{is non-decreasing} \tag{21}$$

when J increases and V belongs to $E_\lambda(I)$. The inequality

$$\Sigma_J : c\,Q_J(U-V, U-V) + (f,f)_{S\,J} \leq (f,f)_{S\,I},\ V \in E_\lambda(I), \tag{22}$$

defines one side of a second order surface in the finite dimensional space $E_\lambda(I)$ or is never satisfied. The sum of the second order terms in (22) is the expression

$$c\underset{J}{Q}(V,V) = c^2(\underset{J}{v},v)_S$$

which is positive definite according to (9) so that (22) is (the interior Σ_J) of an ellipsoid or the empty set. The centre of (22) is determined by an equality which coincides with the formula in (10), if in this formula U(J) is replaced by V. This tells that the centre of (22) is V = U(J). Now the following inequality holds true, viz.

$$c\underset{J}{Q}(U - U(J),\ U - U(J)) \leqslant 0. \tag{23}$$

For the assumption to the contrary implies that $c\underset{J}{Q}$ would be positive definite on the linear hull

$$\{U, E_\lambda(I)\} = \{t(U - U(J)) + V : t = \text{number},\ V \in E_\lambda(I)\}.$$

If U is not in $E_\lambda(I)$ this violates (5). If U belongs to $E_\lambda(I)$, the inequality (23) is trivial since U(J) = U. The inequality (23) shows that Σ_J is non-empty. From (21) it follows that the ellipsoids have the shrinking property that $\Sigma_J \supset \Sigma_{J'}$ when $J \subset J'$. The element U(I) is obtained as the limit of the centre U(J) when $J \to I$. This limit is contained in all ellipsoids and the inequality in (22) is satisfied for all J when V = U(I). Thus (18) is valid. It follows that U - U(I) is in E[I]. The relation (17) expresses the fact that U(I) is the centre of the limit ellipsoid $c\underset{I}{Q}(U - V,\ U - V) \leqslant 0,\ V \in E_\lambda(I)$.

The Theorem is now applied when $U = (u, \dot{u})$ belongs to E[I]. Its condition (15) is then automatically fulfilled. The element U(I) must belong to $E_\lambda[I]$ and is the $\underset{I}{Q}$-projection of U on this space. The inequality (18) shows that $E_\lambda[I]$ is maximal in E[I] with the property that $c\underset{I}{Q}$ is positive definite on this subspace. For if U is outside $E_\lambda[I]$ the difference U - U(I) is $\neq 0$ and gives a non-positive value to $c\underset{I}{Q}(U - U(I),\ U - U(I))$. Thus $c\underset{I}{Q}$ is not positive definite on $\{U, E_\lambda[I]\}$.

In the same way it is seen that $E_{\underline{\lambda}}[I]$ is maximal as a subspace of E[I] on which $c\underset{I}{Q}$ is negative definite. A consequence of these statements about $E_{\lambda}[I]$ and $E_{\bar{\lambda}}[I]$ is that their linear hull can be written as a direct sum

$$F[\lambda,I] = E_{\lambda}[I] \dotplus E_{\bar{\lambda}}[I] \tag{24}$$

and that $F[\lambda,I]$ is a maximal subspace of $E[I]$ with the property that Q_I is non-degenerate on this subspace.

Any finite dimensional subspace F of $E[I]$ which has the same maximal property as $F[\lambda,I]$ is said to be of the type F. For such a space F and every element U in $E[I]$ but outside F, the form Q_I is degenerate on $\{U,F\}$. This means that there exists an element $tU + V \neq 0$, t = number and $V \in F$, such that $Q_I(tU + V, \{U,F\}) = 0$. Since Q_I is non-degenerate on F the number $t \neq 0$ and $U' = -t^{-1}V$ is recognised as the Q_I-projection of U on F. It follows that $Q_I(U-U',U) = 0$ and that $Q_I(U-U', U') = 0$. One obtains $Q_I(U,U) = Q_I(U',U')$ and the corresponding formula $Q_I(U,V) = Q_I(U',V')$ can be deduced by polarization. As a consequence $Q_I(U-U',V) = 0$ holds true for all V in $E[I]$ so that $Q_I(U-U', E[I]) = 0$. The decomposition $U = (U-U') + U'$ is unique and implies the formula

$$E[I] = E[I]^{perp} \dotplus F \tag{25}$$

in which $E[I]^{perp} = \{U \in E[I] : Q_I(U, E[I]) = 0\}$. Clearly all elements in $E[I]^{perp}$ are mapped upon 0 by Q_I-projection on F.

By Q_I-projection on a space of the type F, all elements in an equivalence class $U + E[I]^{perp}$ of the quotient space

$$X = E[I]/E[I]^{perp}$$

are mapped upon one and the same U' in F. It follows that Q_I has a consistent definition on X and that the one-to-one mapping

$$X \rightarrow F \quad (Q_I\text{-projection on } F)$$

is Q-true. Consequently the form Q_I has the same signature on X and on all spaces of the type F, in particular on all spaces $F[\lambda, I]$, λ non-real. Since the signature of cQ_I, $c = i^{-1}(\lambda - \bar{\lambda})$, equals $(\dim E_\lambda[I], \dim E_{\bar{\lambda}}[I])$ on $F[\lambda, I]$, it follows that $\dim E_\lambda[I]$ is constant in the half-plane $\mathrm{Im}(\lambda) > 0$ and also in the half-plane $\mathrm{Im}(\lambda) < 0$. The dimension of $E_\lambda[I]$ equals the number of linearly independent solutions of $Su = \lambda Tu$ which have finite Dirichlet integrals over I.

A Q_I-nullspace Z in E[I] is a linear subspace of E[I] such that

$$Q_I(Z,Z) = 0,$$

i.e. such that $Q_I(U,V) = 0$ for all $U = (u,\dot{u})$ and $V = (v,\dot{v})$ in Z. According to the definition (12) of Q_I an essential property of Z is the validity of the symmetry relation

$$(\dot{u},v)_{S_I} = (u,\dot{v})_{S_I}.$$

Symmetric boundary conditions are maximal Q_I-nullspaces. The purpose of introducing symmetric boundary conditions, here simply called boundary conditions, makes it an advantage (and a necessity) to take them as large as possible. An example of a Q_I-nullspace is $E[I]^{\mathrm{perp}}$. If Z is a Q_I-nullspace, the linear hull $\{Z, E[I]^{\mathrm{perp}}\}$ is also a Q_I-nullspace. As a consequence one can always assume that

$$Z \supset E[I]^{\mathrm{perp}}$$

when Z is a boundary condition. Q_I-projection of a nullspace Z on a space of the type F gives a Q_I-nullspace Z' in F. According to the previous principle Z' should be maximal in F if Z is a boundary condition. This maximality of Z' is equivalent to the condition

$$\dim Z' = \min(\dim E_\lambda[I], \dim E_{\bar{\lambda}}[I] \tag{26}$$

as can be seen by taking $F = F[\lambda, I]$. Corresponding to (25) any boundary condition can be written

$$Z = E[I]^{perp} \dotplus Z',$$

where Z' is a Q_I-nullspace in F satisfying (26).

The previous Theorem has the second consequence that <u>the equation</u> $Su - \lambda Tu = Tf$ <u>for any non-real</u> λ <u>and any</u> f <u>in</u> $C^K(I)$ <u>with</u> $(f,f)_{S_I} < +\infty$, <u>has a solution such that</u> $(u,u)_{S_I} < +\infty$ <u>is satisfied</u>. To obtain this result one first determines a solution u_0 of $Su_0 - \lambda Tu_0 = Tf$ by integration. The element $U_0 = (u_0, \lambda u_0 + f)$ belongs to E(I) and satisfies (15). The solution of the problem is $U_0 - U_0(I)$ where $U_0(I)$ is obtained according to the Theorem.

If λ is taken in the half-plane where

$$\dim E_\lambda[I] \geq \dim E_{\bar\lambda}[I], \tag{27}$$

the solution of $Su - \lambda Tu = Tf$ can be made to satisfy a given boundary condition Z. For let $U_1 = U_0 - U_0(I)$ be the solution of the previous problem so that $U_1 \in E[I]$. Let U_1' be the Q_I-projection of U_1 on $F[\lambda,I]$. Because of (26), (27)

$$F[\lambda,I] = E_\lambda[I] \dotplus Z', \tag{28}$$

where Z' is the image of Z under Q_I-projection on F. Let

$$U_1' = U_2 + W, \quad U_2 \in E_\lambda[I], \quad W \in Z',$$

be the decomposition of U_1' according to (28). Then $U = U_1 - U_2$ has the form $(u, \lambda u + f)$ and $U = (U_1 - U_1') + W \in E[I]^{perp} \dotplus Z' = Z$. It is easy to see that the solution is unique. Let it be denoted by $u = R(\lambda)f$. The equality $u = R(\lambda)f$ is equivalent to $U = (u, \lambda u + f) \in Z$. From $Q_I(U,U) = 0$ it follows, because of the definition (12) and by the help of Cauchy-Schwarz' inequality, that $R(\lambda)$ is bounded with respect to $\|u\|_{S_I} = (u,u)_{S_I}^{1/2}$.

The following conclusions are given without detailed proofs.

Consider the linear set

$$D = \{u : (u,\mathring{u}) \in Z \text{ for a } \mathring{u} \in \overline{Z}_1\}.$$

Here Z_1 consists of all first elements of pairs in Z and $\overline{Z}_1$ is its closure with respect to $\|\cdot\|_{\underset{I}{S}}$. It can be proved that D defines a function $\mathring{u} = Pu$ which is a symmetric and densely defined operator in the Hilbert space $H(Z) = \overline{Z}_1$. This operator is related to the mapping of $R(\lambda)$ in the way that a suitable restriction of $R(\lambda)$ coincides with the resolvent of P for any λ in the half-plane in which $R(\lambda)$ exists. In this half-plane the range of $\overline{P} - \lambda$, where $\overline{P}$ is the closure of P, coincides with the entire Hilbert space $H(Z) = \overline{Z}_1$. Thus $\overline{P}$ is maximal symmetric with one deficiency index 0. The other deficiency index is $\dim E_\lambda[I] - \dim E_{\overline{\lambda}}[I]$. The spectral theory for $\overline{P}$ gives a spectral theory for S and T considered under the restriction of an arbitrary boundary condition Z. The spectral theory takes place in the Hilbert space $H(Z) = \overline{Z}_1$ determined by Z.

In the selfadjoint case, $\dim E_\lambda[I] = \dim E_{\overline{\lambda}}[I]$, it can be proved that eigenspaces $\mathcal{E}(\Delta)$ of $\overline{P}$ belonging to finite intervals Δ consist of regular functions to which the differential operators can be applied. In particular existing eigenfunctions are regular solutions of $Su = \rho Tu$, where ρ is an eigenvalue. The mapping $u = R(\lambda)f$ and its closure can be represented by the help of a Green's function. The representation has the form

$$\lambda u(x) = -f(x) + (f(\cdot), \overline{g(\cdot,x,\lambda)})_{\underset{I}{S}}$$

in a certain contrast to the corresponding formula in [7], where the lower order operator T has a positive Dirichlet integral used as an inner product.

The University, Dundee, Scotland
Uppsala University, Sweden.

References

1. C. Bennewitz and Å. Pleijel, 'Selfadjoint extension of ordinary differential operators.' To appear in the Proceedings of the Colloquium on Mathematical Analysis at Jyväskylä, Finland, August 17-21, 1970.

2. K. Emanuelsson, 'Spectral theory of one formally symmetric ordinary differential operator with respect to another.' Mimeographed preprint, Department of Mathematics, Uppsala University, Sweden, 1 - 16 (1971).

3. T. Kimura and M. Takahasi, ' Sur les opérateurs différentiels ordinaires linéaires formellement autoadjoints I.' Funkcialaj Ekvacioj, Serio Internacia 7, 35-90 (1965).

4. Å. Pleijel, 'Some remarks about the limit point and limit circle theory.' Arkiv för Matematik 7: 41, 543-550 (1968).

5. Å. Pleijel, 'Complementary remarks about the limit point and limit circle theory.' Arkiv för Mathematik 8: 6 45-47 (1969).

6. Å. Pleijel, 'Spectral theory for pairs of ordinary formally self-adjoint differential operators.' Journal of the Indian Mathematical Society 34, 259-268 (1970).

7. Å. Pleijel, 'Green's functions for pairs of formally selfadjoint ordinary differential operators.' Lecture Notes in Mathematics 280, Conference on the Theory of Ordinary and Partial Differential Equations, 131-146 (1972). Springer-Verlag.

8. H. Weyl, 'Über gewöhnliche Differentialgleichungen mit Singularitäten und die zugehörigen Entwicklungen willkürlicher Funktionen.' Math. Ann. 68, 220-269 (1910). Also in Selecta Hermann Weyl, Birkhauser Verlag, Basel and Stuttgart, 1956.

Some remarks on a differential expression with an indefinite weight function

W. N. Everitt

1. This paper is concerned with the generation of a self-adjoint operator, in an integrable-square Hilbert function space, from a boundary value problem associated with the differential equation on a half-line $[0, \infty)$

$$M[y] \equiv -(py')' + qy = \lambda ry \qquad (' \equiv \frac{d}{dx}) \tag{1.1}$$

Here the coefficient p is positive, q is non-negative but the weight function r may take both positive and negative values on $[0, \infty)$; $\lambda = \mu + i\nu$ is a complex parameter.

With these sign conditions on the coefficients p, q and r the differential equation (1.1) is often called the 'polar' equation. Under this name it was studied by Hermann Weyl in 1910; see [10], section 3.

The integrable-square solutions of (1.1) have been considered in recent years by Pleijel and his school at Uppsala. We quote here the two papers of Pleijel [6] and [7]; see also his contribution to this volume [8].

Work on the so-called left-definite systems of differential equations is reported on in the contribution by Niessen and Schneider to this volume; see [4] and, in particular, its list of references.

The operator-theoretic aspects of (1.1), in the case when the equation is considered on the whole real line $(-\infty, \infty)$ and the coefficient r is bounded, are considered in a recent paper [9] by Shotwell.

One form of the 'coefficient of Weyl' for the equation (1.1) is considered by Atkinson, Everitt and Ong in [2] and some of the methods in that work will be adapted for use in this paper. A study of the limit-point/limit-circle classification of Weyl, for the equation (1.1), has been undertaken by Ong in his recent Ph.D. thesis [5]. Both of these references contain information about the integrable-square solutions of (1.1).

Our interests in this paper are similar to those of Shotwell in [9] but we are concerned with an extension to the case when the coefficient r may be unbounded as well as oscillatory. The results of Shotwell are for differential equations of arbitrary even order, defined on the whole real line $(-\infty, \infty)$. Here we consider the case of (1.1) on the half-line $[0, \infty)$ and introduce boundary condition at the regular point 0. We shall show that it is only for certain of these boundary conditions that a self-adjoint operator, associated with (1.1), can be defined. For the other boundary conditions the operator may sometimes be defined but the underlying function space has an indefinite inner product and the ordinary concept of self-adjointness is no longer valid; it is possible that the theory of Iohvidov and Krein in [3] could be applied in such cases but we do not consider such problems in this paper.

We outline the contents of the paper. Section 2 contains preliminary material and section 3 a statement of the results to be proved. The proof of these results takes up sections 4 to 13. The three remaining sections contain some additional material on the case when it is not possible to construct a self-adjoint operator to represent the boundary value problem. There is a list of references.

Acknowledgements I am grateful for helpful discussions with Professor Åke Pleijel and Dr Christer Bennewitz (both of the University of Uppsala, Sweden) during a period when they visited the University of Dundee under the auspices of the Science Research Council of the United Kingdom.

2. We write R for the real line, C for the complex plane, $C_{\pm} = \{\lambda \in C : \operatorname{im} \lambda \gtrless 0\}$ and $C_0 = C_+ \cup C_-$.

The coefficients p, q and r of (1.1) are real-valued on $[0, \infty)$ and satisfy the following basic conditions:

$$p \in AC_{loc}[0, \infty) \quad \underline{\text{and}} \quad p(x) > 0 \qquad (x \in [0, \infty)) \tag{2.1}$$

$$q \in L_{loc}[0, \infty) \quad \underline{\text{and}} \quad q(x) \geq 0 \qquad (x \in [0, \infty)) \tag{2.2}$$

$$r \in L_{loc}[0, \infty) \quad \underline{\text{and}} \quad 0 < \int_0^\infty |r| \leq \infty . \tag{2.3}$$

Here L denotes Lebesque integration, AC absolute continuity and 'loc' a property to be satisfied on all compact subintervals of $[0, \infty)$.

Note that from (2.3) the coefficient r is not identically zero and that there is no restriction of the sign of r on $[0, \infty)$; however r^{-1} may or may not be in $L_{loc}[0, \infty)$.

Further restrictions on the coefficients p, q and r will be given below.

With (2.1, 2 and 3) satisfied it follows that the standard existence theorems may be applied to the differential equation (1.1) to obtain solutions defined on $[0, \infty)$. Thus if the parameter $\alpha \in (-\frac{1}{2}\pi, \frac{1}{2}\pi]$, and θ_α, φ_α denote solutions of (1.1) determined by the following initial conditions at 0

$$\theta_\alpha(0, \lambda) = \cos\alpha \qquad p(0)\theta'_\alpha(0, \lambda) = -\sin\alpha \tag{2.4}$$

$$\varphi_\alpha(0, \lambda) = \sin\alpha \qquad p(0)\varphi'_\alpha(0, \lambda) = \cos\alpha, \tag{2.5}$$

then $\theta_\alpha(\cdot, \lambda)$ and $\varphi_\alpha(\cdot, \lambda)$ are linearly independent solutions of (1.1) on $[0, \infty)$. Further properties of these solutions are given in [2], section 2.

For convenience we write

$$P = [0, \tfrac{1}{2}\pi] \quad N = (-\tfrac{1}{2}\pi, 0) \qquad (-\tfrac{1}{2}\pi, \tfrac{1}{2}\pi] = P \cup N \tag{2.6}$$

and refer to P as the positive, and N the negative values of α.

If the complex-valued function f on $[0, \infty)$ satisfies $f' \in AC_{loc}[0, \infty)$ then the parameter α determines a linear, homogeneous boundary condition for f in the form

$$f(0)\cos\alpha - p(0)f'(0)\sin\alpha = 0. \tag{2.7}$$

The case $\alpha = 0$, $\alpha = \frac{1}{2}\pi$ determine the boundary conditions $f(0) = 0$, $f'(0) = 0$ respectively. All real, linear homogeneous boundary conditions at 0 may be put into the form (2.7), for some $\alpha \in (-\frac{1}{2}\pi, \frac{1}{2}\pi]$, after suitable normalization.

Let $H(= H(p, q : 0, \infty))$ denote the function space of all complex-valued f on $[0, \infty)$ such that $f \in AC_{loc}[0, \infty)$ and

$$\int_0^\infty \{p|f'|^2 + q|f|^2\} < \infty. \tag{2.8}$$

If we endow H with an inner-product

$$(f,\ g)_H = \int_0^\infty \{pf'\overline{g}' + qf\overline{g}\} \qquad (f,\ g \in H) \tag{2.9}$$

then H may be shown to be a Hilbert space. (There are some precautions to be taken if q is identically zero on $[0, \infty)$ but this case will be excluded later.)

For $\alpha \in (-\frac{1}{2}\pi, \frac{1}{2}\pi]$ but $\alpha \neq 0$ let H_α denote all the elements of H but endowed with the scalar product

$$(f,\ g)_\alpha = f(0)\overline{g}(0)\cot\alpha + \int_0^\infty \{pf'\overline{g}' + qf\overline{g}\}. \tag{2.10}$$

The space H_0 is defined by

$$H_0 = \{f \in H : f(0) = 0 \quad (f,\ g)_0 = (f,\ g)_H\}. \tag{2.11}$$

It is not difficult to verify that H_α is a Hilbert space for all $\alpha \in P$; note however that this is not the case when $\alpha \in N$ since then the scalar product (2.10) is, in general, indefinite in sign and so is not an inner-product on H. Note that $(f,\ g)_H = (f,\ g)_{\frac{1}{2}\pi}$ $(f,\ g \in H)$ and so, with a suitable interpretation, $H_{\frac{1}{2}\pi} = H$.

For $\alpha \in P$ we determine a subspace H'_α of H_α, where H'_α depends on the coefficient r, as follows. Let G denote the linear manifold of H defined by

$$G = \{f \in H : rf = 0 \text{ on } [0, \infty), \text{ i.e. } r(x)f(x) = 0 \text{ for almost all } x \in [0, \infty)\}. \tag{2.12}$$

Since, from (2.3), the coefficient r is not the null function on $[0, \infty)$, we see that G is strictly contained in H. Furthermore G consists only of the null vector of H, if and only if

$$\int_a^b |r| > 0 \quad \text{for all compact } [a, b] \subset [0, \infty). \tag{2.13}$$

We now define H'_α as the orthogonal complement of G in H_α, i.e.

$$H'_\alpha = H_\alpha \ominus G. \tag{2.14}$$

Alternatively we may consider H'_α as the quotient space H_α/G.

A classical boundary value problem, denoted here by B_α, on the half-line $[0, \infty)$ may be determined by requiring solutions of the differential equation (1.1) to satisfy the boundary condition (2.7) at 0, and the 'boundary condition' (2.8) at ∞, i.e. the solution should belong to H. In the usual terminology λ is an eigenvalue of this problem B_α if there is a non-trivial solution of the differential equation (1.1) satisfying the boundary conditions (2.7 and 8); this solution is then an eigenfunction associated with the eigenvalue.

When $\alpha \in P$ we shall see that, with some additional conditions on the coefficients p, q and r, we can characterise the problem B_α by means of a uniquely determined unbounded self-adjoint operator T_α in the Hilbert function space H'_α. The eigenvalues and eigenfunctions of B_α are equivalent to the eigenvalues and eigenvectors of T_α in H'_α. These results, however, are not valid when $\alpha \in N$ and we comment again on this point at a later stage.

3. We may now state the result to be proved in this paper in the form of the following

Theorem. Let the real-valued coefficients p, q and r of the differential equation $M[y] = \lambda ry$ on $[0, \infty)$, see (1.1), satisfy the basic conditions (2.1, 2 and 3); additionally let these coefficients satisfy

$$\int_0^\infty p^{-1} = \infty \quad \text{or} \quad \int_0^\infty q = \infty \tag{3.1}$$

and for some positive number K

$$|r(x)| \leqslant Kq(x) \qquad (x \in [0, \infty)); \tag{3.2}$$

then if $\alpha \in P = [0, \frac{1}{2}\pi]$ there is a uniquely determined operator $T_\alpha : D(T_\alpha) \subset H'_\alpha \to H'_\alpha$, with H'_α determined as in section 2, with the properties

(i) the domain $D(T_\alpha) \subset \{f \in H'_\alpha : f' \in AC_{loc}[0, \infty)\}$

(ii) T_α is a self-adjoint, unbounded operator in H'_α

(iii) T_α^{-1} exists and is a bounded operator defined on the whole of H'_α

(iv) for all $f \in H'_\alpha$

$$M[T_\alpha^{-1} f] = rf \text{ almost everywhere on } [0, \infty)$$

(v) T_α represents the boundary value problem B_α in the sense given in section 2.

These results are not valid when $\alpha \in N$.

Proof. This is given in the sections which follow.

It is worth making the following points:

(a) condition (3.2) together with (2.3) implies that q is not identically zero on $[0, \infty)$; this condition was earlier used by Weyl in [10], section 3; see also [2], Theorem 2.

(b) we may replace (3.2) by the more general condition, K again a positive number,

$$\int_0^X |r||f|^2 \leq K \int_0^X \{p|f'|^2 + q|f|^2\} \quad (\text{all } X \geq 0) \tag{3.3}$$

provided this holds with the same K for all $f: [0, \infty) \to C$ such that $f \in AC_{loc}[0, \infty)$; however we work with (3.2) in this paper

(c) we may expect a link between the spectral properties of T_α and the analytic properties of the Weyl type coefficient $m(\cdot, \alpha)$ for (1.1), as defined in [2]

(d) T_α is defined and self-adjoint, and $m(\cdot, \alpha)$ is defined and regular in C_0, both for the same restriction on the boundary condition parameter α, i.e. $\alpha \in P$

(e) the result (iii) of the above theorem shows that the origin of the complex λ plane, the spectral plane, is in the resolvent set of T_α, i.e. not in the spectrum of T_α

(f) a restriction of the form (3.2), or more generally (3.3), is necessary if T_α is to be defined by first constructing the inverse operator T_α^{-1}; we shall give an example to illustrate this point in a later section.

4. Consider now the proof of the above theorem. We adapt the Gram matrix method as given in [2], section 5.

We require the following Lemma; see also the Lemmas given in [2], sections 4 and 7.

Lemma. Let the solution φ_α of (1.1) be determined by the initial conditions (2.5); then for all $\alpha \in P = [0, \frac{1}{2}\pi]$

$$\varphi_\alpha(X, 0) \neq 0 \quad (X > 0) \tag{4.1}$$

and if (3.1) holds

$$\varphi_\alpha(\cdot, 0) \notin H.$$

Proof. This follows the lines of [2], sections 4 and 7 and we omit the details. In both cases we use the identity

$$\int_0^X \{p\,|\varphi_\alpha'|^2 + q\,|\varphi_\alpha|^2\} + \sin\alpha\cos\alpha = p(X)\varphi_\alpha(X, 0)\overline{\varphi}_\alpha'(X, 0)$$

and note that when $\alpha \in P$ we have $\sin\alpha\cos\alpha \geq 0$.

5. In this and following sections we restrict $\alpha \in P$ until the proof of the Theorem is completed.

We now construct a unique solution $\Phi_\alpha(x; f)$, defined for all $x \in [0, \infty)$ and for all $f \in H_\alpha$, of the non-homogeneous differential equation

$$M[\Phi_\alpha] = rf \text{ on } [0, \infty) \tag{5.1}$$

such that

(i) $\Phi_\alpha(\cdot\,; f)$ satisfies the boundary condition (2.7) at 0

(ii) $\Phi_\alpha(\cdot\,; f) \in H_\alpha$.

Given any $f \in H_\alpha$ define the solution $\chi_\alpha(\cdot\,; f)$ by

$$\chi_\alpha(x; f) = \theta_\alpha(x, 0)\int_0^x \varphi_\alpha(\cdot, 0)rf - \varphi_\alpha(x, 0)\int_0^x \theta_\alpha(\cdot, 0)rf \quad (x \in [0, \infty)) \tag{5.2}$$

It may be verified that

$$M[\chi_\alpha] = rf \quad \text{on } [0, \infty) \tag{5.3}$$

and that

$$\chi_\alpha(0; f) = \chi'_\alpha(0; f) = 0. \tag{5.4}$$

Given $X > 0$ define the solution $\Phi_{\alpha,X}(\cdot\,; f)$ on $[0, X]$ by

$$\Phi_{\alpha,X}(x; f) = \gamma_\alpha(X; f)\varphi_\alpha(x, 0) + \chi_\alpha(x; f) \quad (x \in [0, X]) \tag{5.5}$$

choosing $\gamma_\alpha(X; f) \in C$ so that

$$\Phi_{\alpha,X}(X; f) = 0 \tag{5.6}$$

i.e.

$$\gamma_\alpha(X; f) = -\chi_\alpha(X; f)\{\varphi_\alpha(X, 0)\}^{-1}. \tag{5.7}$$

It follows from (4.1) that γ_α is well defined on $(0, \infty) \times H_\alpha$. We note that $M[\Phi_{\alpha,X}] = rf$ on $[0, X]$; also $\Phi_{\alpha,X}$ satisfies the boundary condition (2.7) at 0.

6. In the space H_α, when $\alpha \in P$ but $\alpha \neq 0$, we have the inner-product $(f, g)_\alpha$, see (2.10), and the corresponding norm $\|f\|_\alpha$. It is convenient to introduce the restricted inner-product, defined for all $f, g \in AC_{loc}[0, X]$,

$$(f, g)_{\alpha,X} = \int_0^X \{pf'\overline{g}' + qf\overline{g}\} + f(0)\overline{g}(0)\cot\alpha \tag{6.1}$$

with a corresponding norm $\|f\|_{\alpha,X}$.

Similarly $(f, g)_{0,X}$ and $\|f\|_{0,X}$ are defined in such a way as to be consistent with (2.11).

7. Consider for the moment $\alpha \in P$ but $\alpha \neq 0$.

Integration by parts yields the identity

$$\int_0^X \{p|\Phi'_{\alpha,X}|^2 + q|\Phi_{\alpha,X}|^2\} = [p\Phi_{\alpha,X}\overline{\Phi}'_{\alpha,X}]_0^X + \int_0^X \Phi_{\alpha,X}\overline{M[\Phi_{\alpha,X}]}.$$

Using (5.6) and the other properties of $\Phi_{\alpha,X}$ given in section 5, this last result may be re-written as

$$\|\Phi_{\alpha,X}\|^2_{\alpha,X} = \int_0^X \Phi_{\alpha,X} r \bar{f} \tag{7.1}$$

valid for $f \in H_\alpha$ and all $X > 0$.

In (7.1) we now make use of the condition (3.2) or r and q to obtain

$$\begin{aligned}\|\Phi_{\alpha,X}\|^2_{\alpha,X} &\leqslant K\int_0^X q\,|\Phi_{\alpha,X} f\,| \\ &\leqslant K\left\{\int_0^X q\,|\Phi_{\alpha,X}|^2 \int_0^X q\,|f\,|^2\right\}^{\frac{1}{2}} \\ &\leqslant K\,\|\Phi_{\alpha,X}\|_{\alpha,X}\;\|f\,\|_{\alpha,X}\end{aligned}$$

and so

$$\|\Phi_{\alpha,X}\|_{\alpha,X} \leqslant K\|f\,\|_{\alpha,X} \quad (X > 0,\ f \in H_\alpha). \tag{7.2}$$

A separate calculation shows that (7.2) holds also in the case $\alpha = 0$, noting that $\Phi_{0,X}(0\,;\,f) = 0$.

8. Suppose now $f \in H'_\alpha$, as defined in (2.4), but $f \not\equiv 0$; then rf is not null on $[0, \infty)$ and so, from the equation (5.3), $\chi_\alpha(\cdot\,;\,f)$ is not null on $[0, X]$ when X is sufficiently large, say $X \geqslant X_0 = X_0(f)$.

From this requirement on f it follows that φ_α and χ_α are then linearly independent on $[0, X]$ for all $X \geqslant X_0$. For otherwise we should have a number $A \in C$ for which $A\varphi_\alpha + \chi_\alpha = 0$ on $[0, X_0]$; if $0 < \alpha \leqslant \frac{1}{2}\pi$ then taking $x = 0$ we obtain, from (2.5) and (5.4), $A = 0$; if $\alpha = 0$ then differentiating and taking $x = 0$ we again obtain $A = 0$. However this is a contradiction since χ_α is not null on $[0, X_0]$.

Consider now the Gram matrix, see [2], section 5, again for $f \in H'_\alpha$

$$\Gamma(X, f) = \begin{bmatrix} \int_0^X q\,|\varphi_\alpha|^2 & \int_0^X q\,\varphi_\alpha \overline{\chi}_\alpha \\ \int_0^X q\,\chi_\alpha \overline{\varphi}_\alpha & \int_0^X q\,|\chi_\alpha|^2 \end{bmatrix}$$

From (2.3) and (3.2) we have $\int_0^X q > 0$ when X is sufficiently large, say $X \geqslant X_1$. From this and the linear independence of φ_α and χ_α we see that $\Gamma(X; f)$ is positive definite when $X \geqslant \max\{X_0(f), X_1\} = X_3(f)$ (say).

Let $\kappa(X; f)$ denote the smallest characteristic root of $\Gamma(X; f)$ then

(i) $\kappa(X; f) > 0 \quad (X \geqslant X_3(f))$

(ii) $\kappa(X; f) \leqslant \kappa(X'; f) \quad (X' > X \geqslant X_3(f))$.

For (ii) see [2], section 5 and the references given.

9. Since $\alpha \in P$ we obtain from (7.2) the inequality

$$\int_0^X q\,|y_\alpha(X; f)\varphi_\alpha + \chi_\alpha|^2 \leqslant K^2 \|f\|^2_{\alpha,X} \quad (X > 0). \tag{9.1}$$

The left-hand side of this inequality is a positive definite form with matrix $\Gamma(X; f)$.

If $f \in H'_\alpha$, with $f \neq 0$, then following the method in [2], section 5 (in particular (5.3)), we obtain the following inequality for $y_\alpha(X; f)$, using (ii) of the previous section,

$$\begin{aligned} |y(X; f)|^2 &\leqslant K^2\{\kappa(X; f)\}^{-1} \|f\|^2_{\alpha,X} \\ &\leqslant K^2\{\kappa(X_3(f); f)\}^{-1} \|f\|^2_\alpha \end{aligned} \tag{9.2}$$

valid for all $X \geqslant X_3(f)$.

If $f \notin H'_\alpha$, i.e. $f \in G$ as defined by (1.12), then rf is null, $\chi_\alpha(\cdot; f)$ is null and $y_\alpha(X; f) = 0$ for all $X > 0$.

Thus for all $f \in H_\alpha$ we have a positive number $L(f)$, depending on f, such that

$$|y_\alpha(X; f)| \leqslant L(f) < \infty \tag{9.3}$$

holds for all sufficiently large X. This result implies the existence of a sequence $\{X_n; n = 1, 2, \dots\}$, depending on f, such that $\lim_{n\to\infty} X_n = \infty$ and $\lim_{n\to\infty} y_\alpha(X_n; f)$ exists and is finite, say $\delta_\alpha(f)$.

We now define $\Phi_\alpha(\cdot; f)$ on $[0, \infty)$ by

$$\Phi_\alpha(x; f) = \delta_\alpha(f)\varphi_\alpha(x, 0) + \chi_\alpha(x; f) \quad (x \in [0, \infty)) \tag{9.4}$$

and note that

$$\lim_{n\to\infty} \Phi_{\alpha,X_n}(\cdot; f) = \Phi_\alpha(\cdot; f) \text{ on } [0, \infty) \tag{9.5}$$

with bounded convergence on any interval $[0, X]$.

We also obtain

$$\lim_{n\to\infty} \Phi'_{\alpha,X_n}(\cdot; f) = \Phi'_\alpha(\cdot; f) \text{ on } [0, \infty) \tag{9.6}$$

again with bounded convergence on any interval $[0, X]$.

Given any $X > 0$ we obtain from the inequality (7.2), for all n such that $X_n > X$,

$$\|\Phi_{\alpha,X_n}\|_{\alpha,X} \leqslant \|\Phi_{\alpha,X_n}\|_{\alpha,X_n} \leqslant K\|f\|_{\alpha,X_n}.$$

Now let $n \to \infty$; the bounded convergence of (9.5 and 6) on $[0, X]$ gives

$$\|\Phi_\alpha\|_{\alpha,X} \leqslant K\|f\|_\alpha \qquad (X > 0);$$

in this last result let $X \to \infty$ to obtain

$$\|\Phi_\alpha\|_\alpha \leqslant K\|f\|_\alpha \qquad (f \in H_\alpha) \tag{9.7}$$

noting that this result is valid for all $f \in H_\alpha$.

From the definition (9.4) we see that Φ_α satisfies the non-homogeneous equation (5.1) and the boundary condition (2.7); from (9.7) we see that $\Phi_\alpha \in H_\alpha$ for all $f \in H_\alpha$.

10. Consider now the uniqueness of Φ_α.

Suppose that in the definition of $\delta_\alpha(f)$ the bounded family of numbers $\{y_\alpha(X; f); X > 0\}$ has more than one limit-point, say $\delta_\alpha(f)$ and $\epsilon_\alpha(f)$ with $\delta_\alpha(f) \neq \epsilon_\alpha(f)$; then on substracting the respective Φ_α we find $\varphi_\alpha(\cdot, 0) \in H$ and this is a contradiction on the result of the Lemma in section 4 above. Thus $\{y_\alpha(X; f); X > 0\}$ has a unique limit-point and it is, in fact, not necessary to consider a sequential limit, i.e. we may let X tend continuously to infinity; so

$$\lim_{X\to\infty} y_\alpha(X; f) = \delta_\alpha(f) \quad (f \in H_\alpha). \tag{10.1}$$

This argument also establishes the uniqueness of the solution $\Phi_\alpha(\cdot; f)$ for each $f \in H_\alpha$.

11. We now define a bounded, linear symmetric operator R_α on the whole of H_α by

$$R_\alpha f = \Phi_\alpha(\cdot; f) \quad (f \in H_\alpha). \tag{11.1}$$

From the definition of $y_\alpha(X; f)$ in (5.7) we see that $y_\alpha(X; \cdot)$ is a linear operator on H_α; since the limit definition of $\delta_\alpha(\cdot)$ in (10.1) is the same (i.e. it is not a sequential limit) for all f we see that $\delta_\alpha(\cdot)$ is also a linear operator on H_α. From the definition of Φ_α in (9.4) it now follows that R_α is a linear operator on H_α.

The boundedness of R_α on H_α follows from the inequality (9.7) above.

It remains to prove that R_α is symmetric in H_α. Let $f, g \in H_\alpha$; as in the proof of (7.1) integration by parts establishes the identity

$$\begin{aligned}(R_\alpha f, g)_{\alpha,X} &= (\Phi_\alpha(\cdot; f), g)_{\alpha,X} \\ &= p(X)\Phi'_\alpha(X; f)\bar{g}(X) + \int_0^X r f \bar{g}\,. \end{aligned} \tag{11.2}$$

From (3.2) and $f, g \in H_\alpha$ it follows that the integral on the right of (11.2) is absolutely convergent; this implies that

$$\lim_{X\to\infty} p(X)\Phi'_\alpha(X; f)\bar{g}(X) \tag{11.3}$$

exists and is finite. If this limit is not zero then we may proceed as in [2], section 9 to obtain a contradiction. Indeed with (11.3) not zero it may be shown that for some $X_0 > 0$ we have $g(X) \neq 0$ $(X \in [X_0, \infty))$, $\{p^{\frac{1}{2}}g(\cdot)\}^{-1} \in L(X_0, \infty)$ and that $\lim_{X\to\infty} g(X)$ exists, is finite and not zero. This result, however, is impossible in view of the condition (3.1) on the coefficients p and q.

Thus from (11.2 and 3) we obtain

$$(R_\alpha f, g)_\alpha = \int_0^\infty rf\bar{g} \qquad (f, g \in H_\alpha)$$

and with a similar argument

$$(f, R_\alpha g)_\alpha = \int_0^\infty rf\bar{g} \qquad (f, g \in H_\alpha)$$

and this establishes the symmetry of R_α.

Since R_α is bounded on H_α it follows that R_α is self-adjoint; it follows also that R_α is bounded and self-adjoint in H'_α.

12. To obtain the self-adjoint operator T_α of the Theorem we have to consider the inverse operator R_α^{-1}. In general this inverse operator does not exist in H_α since, as we have seen in section 9, $R_\alpha f = \Phi_\alpha(\cdot\,; f) = 0$ in H_α whenever $f \in G$, as defined in (1.12). However we do have the result that $R_\alpha f = 0$ in H'_α if and only if f is the null vector of H'_α, i.e. the null vector of H_α, as we now establish.

Clearly if f is the null vector then $R_\alpha f = 0$.

Suppose then $R_\alpha f = 0$ for some $f \in H'_\alpha$; then $\Phi_\alpha(\cdot\,; f)$ is null on $[0, \infty)$ and so, from the equation (5.1)

$$0 = M[\Phi_\alpha] = rf \text{ on } [0, \infty)$$

which implies that f is the null vector in H'_α.

Thus R_α^{-1} exists in H'_α.

13. We now define the operator T_α of the Theorem in section 3 by

$$T_\alpha = R_\alpha^{-1} \text{ in } H'_\alpha. \tag{13.1}$$

Consider now the results stated in the Theorem.

(i) The domain $D(T_\alpha)$ is the range of R_α in H'_α and all such vectors are of the form $\Phi_\alpha(\cdot\,; f)$, for some $f \in H'_\alpha$, and so are solutions of (5.1); thus $D(T_\alpha) \subset \{f \in H'_\alpha : f' \in AC_{loc}[0, \infty)\}$.

(ii) Since R_α is bounded and self-adjoint in H'_α its inverse R_α^{-1} is self-adjoint; see [1], section 41 (Corollary to Theorem 1). In view of (i) the domain $D(T_\alpha)$ cannot be the whole of H'_α; R_α^{-1} is closed; if R_α^{-1} were bounded it would have an extension by closure and this gives a contradiction. Thus $T_\alpha = R_\alpha^{-1}$ is unbounded in H'_α.

(iii) $T_\alpha^{-1} = (R_\alpha^{-1})^{-1} = R_\alpha$ and so is a bounded operator defined on the whole of H'_α.

(iv) $T_\alpha^{-1}f = R_\alpha f = \Phi_\alpha(\cdot\,; f)$ and so from (5.1) $M[T_\alpha^{-1}f] = rf$ almost everywhere on $[0, \infty)$; this holds for all $f \in H'_\alpha$.

(v) Suppose that λ is an eigenvalue and ψ a corresponding eigenvector of T_α, i.e. $T_\alpha\psi = \lambda\psi$; then λ is real and $\psi = R_\alpha(T_\alpha\psi) = \lambda R_\alpha\psi$; thus the differential expression $M[\cdot]$ may be applied to ψ and we obtain $M[\psi] = \lambda M[R_\alpha\psi] = \lambda M[\Phi_\alpha(\cdot\,; \psi)] = \lambda r\psi$ on $[0, \infty)$; ψ satisfies the boundary condition at 0 and $\psi \in H'_\alpha$, i.e. $\psi \in H$. Thus λ is an eigenvalue and ψ a corresponding eigenfunction of B_α.

On the other hand suppose λ is an eigenvalue and ψ a corresponding eigenfunction of B_α, i.e. $\psi \in H$; then $\psi \in H_\alpha$ and $M[\psi] = \lambda r\psi$: also $M[\lambda R_\alpha\psi] = \lambda M[\Phi_\alpha(\cdot\,; \psi)] = \lambda r\psi$; thus $M[\psi - \lambda R_\alpha\psi] = 0$ and from the Lemma of section 4 above $\psi - \lambda R_\alpha\psi$ is the null function on $[0, \infty)$ since it also satisfies the boundary condition at 0; it follows that $T_\alpha\psi = \lambda T_\alpha R_\alpha\psi = \lambda\psi$ and so λ is an eigenvalue and ψ a corresponding eigenvector of T_α; note, in particular, that λ is real.

This concludes our discussion of the Theorem when $\alpha \in P$.

14. When $\alpha \in N$ it is not possible to apply the above ideas and methods. Firstly the spaces H_α have an indefinite inner-product and it is not

possible to consider the usual theory of symmetric and self-adjoint operators. Secondly the possible existence of complex eigenvalues for the boundary value problem B_α is indicated in [2], see the results stated in section 3, and any description of the operators associated with B_α would have to take this into account.

15. We discuss briefly one example which shows that if the self-adjoint operator T_α is to be defined as the inverse of R_α then some restriction on the growth at infinity of the coefficient r is essential. We do this by considering an example for which the construction of the solution Φ_α, see section 5 above, is impossible for some f in H_α.

Let

$$p(x) = q(x) = 1 \ \underline{\text{and}} \ r(x) = (x + 1)^{\tau} \quad (x \in [0, \infty))$$

where $\tau > \frac{1}{2}$, i.e. the non-homogeneous equation (5.1) takes the form

$$- \Phi_\alpha'' + \Phi_\alpha = rf \ \text{ on } [0, \infty). \tag{15.1}$$

Consider f defined on $[0, \infty)$ by $f(x) = (x + 1)^{-\tau}$; it may be seen that $f \in H_\alpha$ in this case. The equation (15.1) becomes

$$- \Phi_\alpha'' + \Phi_\alpha = 1 \ \text{ on } [0, \infty)$$

for which the general solution is (A and B arbitrary constants)

$$\Phi_\alpha(x) = Ae^{x} + Be^{-x} + 1 \quad (x \in [0, \infty));$$

all such solutions have the property that $\Phi_\alpha \notin H_\alpha$.

We note that in this example the condition (3.1) is not satisfied.

16. It is possible to construct the operator T_α by other methods and it is hoped to give consideration to such problems in a subsequent paper.

References

1. N. I. Akhiezer and I. M. Glazman, Theory of linear operators in Hilbert space (Ungar, New York, 1961; translated from the Russian edition).

2. F. V. Atkinson, W. N. Everitt and K. S. Ong, 'On the m-coefficient of Weyl for a differential equation with an indefinite weight function,' (to appear in Proc. Lond. Math. Soc.).

3. I. S. Iohvidov and M. G. Krein, 'Spectral theory of operators in spaces with an indefinite metric; Parts I and II,' Amer. Math. Soc. Trans. 13, 105-175 and 34, 283-373.

4. H-D. Niessen and A Schneider, 'Spectral theory for left-definite singular systems of differential equations,' (to be published; see also these Proceedings page 29).

5. K. S. Ong, The limit-point and limit-circle theory of second-order differential equations with an indefinite weight function, Ph.D. thesis, University of Toronto, Canada, 1973).

6. Å. Pleijel, 'Some remarks about the limit point and limit circle theory,' Arkiv för Matematik 7 (1969) 543-550.

7. Å. Pleijel, 'Complementary remarks about the limit point and limit circle theory,' Arkiv för Matematik 8 (1971) 45-47.

8. Å. Pleijel, 'Boundary conditions for pairs of self-adjoint differential operators,' (to be published; see also these Proceedings page 1).

9. D. A. Shotwell, 'Singular boundary value problems for the differential equation $Lu = \lambda \sigma u$, ' Rocky Mountain J. of Math. 1 (1971) 687-708.

10. Hermann Weyl, 'Über gewöhnliche Differentialgleichungen mit singulären Stellen und ihre Eigenfunktionen (2. Note),' Nachrichten der Königlichen Gesellschaft der Wissenschaften zu Göttingen. Mathematisch-physikalische Klasse (1910) 442-467.

Department of Mathematics
The University
DUNDEE
Scotland, U.K.

SPECTRAL THEORY FOR LEFT-DEFINITE SINGULAR SYSTEMS OF DIFFERENTIAL EQUATIONS I

by H.D. NIESSEN and A. SCHNEIDER

ABSTRACT: Singular left-definite systems of differential equations and the corresponding boundary-value problems are considered. The spectral theory of such systems is deduced; especially a norm-expansion theorem and a direct expansion theorem are derived.

1. INTRODUCTION

In 1910 H. Weyl [8], [9] considered the differential equation

$$(1.1)\quad -(p\eta')' + q\eta = \lambda r\eta$$

on the half-axis $[0,\infty)$. He assumed that the coefficients p,q and r are real-valued, continuous and that p is positive. Furthermore, he made the assumption that

$$(1.2)\quad r(x) > 0 \text{ on } [0,\infty)$$

or that there exists a positive constant ρ such that

$$(1.3)\quad |r(x)| \leqq \rho q(x) \text{ on } [0,\infty).$$

In case of (1.2) the equation (1.1) is called "right-definite", in case of (1.3) it is called "left-definite" or "polar". In both cases Weyl considered boundary-value problems arising from (1.1) and proved (direct) expansion theorems belonging to solutions of (1.1).

Since 1910 a lot of papers have been published concerning the right-definite case and generalizations of it to differential equations of higher order and to systems of differential equations. For references compare e.g. [1] and the literature given there.

On the other hand the left-definite case has not been treated for a long time. The first treatment in the direction of singular left-definite differential equations of higher order known to the authors is the thesis of Shotwell [7]. Like Weyl, Shotwell assumes that the equation considered is in the limit-point case, i.e., boundary conditions have only to be imposed at the regular endpoint 0 of the interval $[0,\infty)$.

In this paper we do not need this assumption. Furthermore, we consider left-definite systems of differential equations. As is shown in the second part of this paper, the equations considered by Weyl and Shotwell are special cases of such systems. Moreover, the boundary conditions are allowed to depend linearly on the eigenvalue parameter. It seems that until now no attempt has been made to prove expansion theorems for such singular left-definite systems of differential equations. In the regular case, i.e., in case of a compact interval, left-definite systems have been treated by Schäfke and A. Schneider [3],[4],[5].

Complete proofs of the results given here and further results on singular left-definite differential systems and the corresponding boundary-value problems will be published elsewhere [6].

2. THE PROBLEM

We consider systems of differential equations of the form

(2.1) $F_{11}y' + F_{12}y = \lambda G_1 y$

on an arbitrary interval I with endpoints a and b (a<b) which are allowed to be $-\infty$ and ∞, resp. We make the following assumptions:

(1) There exist $k \in \mathbb{N}$ and continuous mappings D_{11}, D_{12}, D_{22}, E_{11} from I into the set of all complex (k,k) - matrices such that

$$F_{11}(x) = E_{2k},\ F_{12}(x) = \begin{pmatrix} -D_{12}^*(x) & -D_{22}(x) \\ D_{11}(x) & D_{12}(x) \end{pmatrix},\ G_1(x) = \begin{pmatrix} 0 & 0 \\ E_{11}(x) & 0 \end{pmatrix}$$

for $x \in I$. Here E_{2k} denotes the identity-matrix of order 2k.

(2) $D_{11}(x)$, $D_{22}(x)$ and $E_{11}(x)$ are hermitian for all $x \in I$.

With these assumptions the system (2.1) is called "canonical". Such canonical systems arise e.g. in the Calculus of Variation.

Let us abbreviate the differential operator on the left

side of (2.1) by F_1:

$$F_1 y := F_{11} y' + F_{12} y.$$

Then we suppose that the system (2.1) is left-definite, which means that the following three conditions are fulfilled:

(3) $D_{11}(x) \geq 0 \geq D_{22}(x)$ for $x \in I$,

(4) For some constant $\rho > 0$ and all $x \in I$
$$-\rho D_{11}(x) \leq E_{11}(x) \leq \rho D_{11}(x)$$

(5) $F_1 y = 0$ and $\int_I y^*(x) \begin{pmatrix} D_{11}(x) & 0 \\ 0 & -D_{22}(x) \end{pmatrix} y(x)dx = 0$ imply $y = 0$.

For later use we define

$$(2.2) \quad S_1 := \begin{pmatrix} 0 & -E_k \\ E_k & 0 \end{pmatrix}$$

We want to define boundary conditions which together with equation (2.1) give rise to a selfadjoint operator in some Hilbert space. Then the spectral measure belonging to this operator has to be calculated. If this is done the spectral theorem implies a norm-expansion theorem connected with solutions of (2.1).

To define boundary conditions we have to introduce

3. SOME SPACES

First let $\mathfrak{B}$ be the space of all measurable functions defined almost everywhere on I with values in $\mathbb{C}^{2k}$. Equality in $\mathfrak{B}$ means equality almost everywhere. Furthermore, let $\mathfrak{A}$ be that subspace of $\mathfrak{B}$ consisting of all functions which are defined and locally absolutely continuous on all of I. Then the differential operator F_1 is defined on $\mathfrak{A}$ and maps $\mathfrak{A}$ linearly into $\mathfrak{B}$:

$$F_1 : \mathfrak{A} \to \mathfrak{B} \text{ lin.}$$

Also, the multiplication of functions in $\mathfrak{B}$ by the matrix-valued function G_1 or by the matrix S_1 are linear mappings from $\mathcal{L}$ into itself which will be denoted by the same symbols G_1 and S_1, resp.

By assumption (3) $D_{11}(x)$ and $-D_{22}(x)$ are positive-semidefinite. Therefore there exists a positive-semidefinite square-root $U_1(x)$ of $\begin{pmatrix} D_{11}(x) & 0 \\ 0 & -D_{22}(x) \end{pmatrix}$:

$$(3.1)\quad U_1^2(x) = \begin{pmatrix} D_{11}(x) & 0 \\ 0 & -D_{22}(x) \end{pmatrix}.$$

Since D_{11} and D_{22} are continuous, the same is true for U_1. Like G_1 and S_1 we consider multiplication by U_1 as a linear mapping from $\mathcal{L}$ into itself, denoted by U_1, too.

Now let J be an arbitrary subinterval of I and denote by $L^2(J)$ the set of all functions y in $\mathcal{L}$, for which y^*y is integrable over J. Then E(J) is defined to be the inverse image of $L^2(J)$ with respect to U_1:

$$E(J) := U_1^{-1}L^2(J).$$

$E_\lambda(J)$ shall denote the space of solutions of (2.1) lying in E(J):

$$E_\lambda(J) := \{y \in E(J) \cap \mathfrak{A} \mid F_1 y = \lambda G_1 y\}.$$

Furthermore, define the linear space R by

$$R := E(I) \cap F_1^{-1}(G_1 E(I)).$$

Then the following inclusions are true

$$(3.2)\quad \begin{array}{ccc} & \mathfrak{A} \subset \mathcal{L} & \\ & \cup \quad\; \cup & \\ E_\lambda(I) \subset & R \subset E(I). & \end{array}$$

For $y,z \in \mathcal{L}$ with z^*y integrable over J define

$$[y,z]_J := \int_J z^*(x)y(x)dx.$$

Then especially $[y,z]_J$ is an inner product[1] on $L^2(J)$ and we may define an inner product on $E(J)$ by mapping elements $y,z\in E(J)$ by U_1 into $L^2(J)$ and taking the inner product there:

$$(y,z)_J := [U_1y, U_1z]_J \quad (y,z\in E(J)).$$

Then (3.1) implies

$$(3.3)\quad (y,z)_J = \int_J z^*(x)\begin{pmatrix} D_{11}(x) & 0 \\ 0 & -D_{22}(x)\end{pmatrix} y(x)dx \qquad (y,z\in E(J)).$$

This shows that assumption (5) guaranties the positive-definiteness of $(\ ,\ )_I$ on the "eigenspace" $E_o(I)$.

The norms induced by these inner products will be denoted by

$$||y||_J := (y,y)_J^{1/2} \quad (y\in E(J)).$$

Then it can be shown that $(E(J), ||\ ||_J)$ is complete.

We shall prove the following

(3.4) Lemma: For all $y,z\in E(J)$ $(S_1z)^*G_1y$ is integrable over J and

$$|[G_1y, S_1z]_J| \le 4\rho ||y||_J ||z||_J.$$

Proof: By (2.2) and assumptions (1),(3) and (4) we get for $y\in E(J)$

$$|(S_1y)^*G_1y| = |y^*\begin{pmatrix} E_{11} & 0 \\ 0 & 0\end{pmatrix} y| \le \rho y^*\begin{pmatrix} D_{11} & 0 \\ 0 & -D_{22}\end{pmatrix} y$$

and therefore (3.3) shows that $(S_1y)^*G_1y$ is integrable and that

$$|[G_1y, S_1y]_J| \le \rho\ ||y||_J^2.$$

Now the lemma follows by using polar decomposition.

4. THE EIGENSPACES $E_o(I_a)$, $E_o(I_b)$

Let $x_o\in I$ and define $I_a := I\cap[a,x_o], I_b := I\cap[x_o,b]$. We consider $E_o(I_a)$ and $E_o(I_b)$, i.e., the spaces of solutions of

1) In this note "inner product" always means a positive-semi-definite, not necessarily definite hermitian sesquilinear form.

$F_1 y = 0$ which are "square-integrable" at a and b resp. in the sense that $||y||_{I_a}$ and $||y||_{I_b}$ resp. are finite. Since the elements of $E_\lambda(J)$ are continuous, $E_o(I_a)$ and $E_o(I_b)$ do not depend on the choice of x_o. To investigate these spaces denote by Y a fundamental matrix of solutions of $F_1 y = 0$ and define

$$D: = -Y^*\begin{pmatrix} 0 & E_k \\ E_k & 0 \end{pmatrix} Y.$$

Then D(x) is hermitian, regular and has k positive and k negative eigenvalues. Furthermore, it can be shown that the Gram-matrix $(Y,Y)_J$ of columns of Y satisfies the equality

$$(4.1) \quad (Y,Y)_J = \tfrac{1}{2}(D(\beta) - D(\alpha))$$

for each compact interval $J = [\alpha,\beta] \subset I$. Therefore, D is monotonously increasing. Furthermore, it follows from (4.1) that

$$(4.2) \quad Yf \in E_o(I_a) \quad \text{iff} \quad \text{ex.} \lim_{\alpha \to a} f^* D(\alpha) f.$$

Thus, $E_o(I_a)$ can be characterized by limits of eigenvectors of D:

Let

$$(4.3) \quad \mu_1(x) \geq \dots\dots \geq \mu_k(x) > 0 > \mu_{k+1}(x) \geq \dots\dots \geq \mu_{2k}(x)$$

be the eigenvalues of D(x) and denote by

$$(4.4) \quad v_1(x), \dots\dots, v_{2k}(x)$$

an orthonormal system of corresponding eigenvectors. Since D is monotonously increasing, so are the eigenvalues and therefore the following limits exist, possibly being $-\infty$:

$$(4.5) \quad \mu_j: = \lim_{x \to a} \mu_j(x).$$

In view of (4.3) these limits decrease with increasing j. Define σ by

$$\mu_1 \geq \mu_2 \geq \dots \geq \mu_\sigma > -\infty = \mu_{\sigma+1} = \dots = \mu_{2k}.$$

Then (4.3) and (4.5) imply

(4.6) $\sigma \geq k$.

Since the set of eigenvectors in (4.4) is orthonormal, there exists a sequence x_μ converging to a, such that the following limits exist:

$$v_j := \lim_{\mu\to\infty} v_j(x_\mu).$$

Obviously, these limits form an orthonormal system, too. Then it follows from [2], Lemma 3.1

$$\lim_{\alpha\to a} f^* D(\alpha) f \text{ exists iff } f \in \text{span } \{v_1,..,v_\sigma\}.$$

Therefore (4.2) proves the first part of the following

(4.7) Theorem: $E_o(I_a) = \{Yf \mid f \in \text{span } \{v_1,...,v_\sigma\}\}$,

$E_o(I_b) = \{Yf \mid f \in \text{span } \{w_{\nu+1},...,w_{2k}\}\}$.

The second equality can be proved similarly. In this case

$$w_j = \lim_{\mu\to\infty} v_j(y_\mu)$$

for a certain sequence y_μ converging to b, and ν is maximal with

$$\lim_{x\to b} \mu_\nu(x) = \infty$$

By (4.3) we get

(4.8) $\nu \leq k$.

Using (4.8) and (4.6) we obtain the following

(4.9) Corollary: $\dim E_o(I_a) = \sigma \geq k$,

$\dim E_o(I_b) = 2k - \nu \geq k$.

Especially, there are at least k solutions of $F_1 y = 0$ which are "square-integrable" at a and k solutions "square-integrable" in a neighbourhood of b.

Furthermore, it is possible to show that $\mathbb{C}^{2k}$ is spanned by $v_1,...,v_k$, $w_{k+1},...,w_{2k}$:

(4.10) $\mathbb{C}^{2k} = \text{span } \{v_1,...v_k\} \dotplus \text{span } \{w_{k+1},..., w_{2k}\}$.

In connection with Theorem (4.7) this e.g. implies that each solution of $F_1 y = 0$ can be decomposed into the sum of two solutions, one being "square-integrable" at a, the other at b:

$$Y\mathbb{C}^{2k} = E_o(I_a) + E_o(I_b).$$

Denote the projections of $\mathbb{C}^{2k}$ onto span $\{v_1,\dots,v_k\}$ and span $\{w_{k+1},\dots,w_{2k}\}$ according to the decomposition (4.10) by P_a and P_b, resp. Then, for $y \in E(I)$ and $x \in I$ we define

$$(4.11)\quad (A_1 y)(x) := Y(x)\{P_b S_1 \int_a^x (S_1 Y P_a)^*(t)(G_1 y)(t)dt - P_a S_1 \int_x^b (S_1 Y P_b)^*(t)(G_1 y)(t)dt\}.$$

The integrals exist: By definition of P_a and by theorem 4.7 the columns of YP_a ly in $E_o(I_a)$. But since $E_o(I_a)$ does not depend on x_o, they are contained in $E_o(I \cap [a,x])$. The same is true for y. Therefore, lemma 3.4 shows that the first integral in (4.11) exists and depends continuously on y. The same is true for the second integral in (4.11).

Using some other properties of P_a and P_b the following theorem can be proved:

<u>(4.12) Theorem:</u> $A_1 : E(I) \longrightarrow R$ lin., cont.

$$F_1 A_1 y = G_1 y \text{ for } y \in E(I).$$

Especially, $A_1 y$ is a "square-integrable" solution z of the inhomogeneous equation $F_1 z = G_1 y$.

Since R is contained in $\mathfrak{a}$ (compare (3.2)), it consists of continuous functions. Therefore, on R we may consider the topology τ of local uniform convergence. Then, since the integrals in (4.11) and hence the expression in brackets depend continuously on y and x, A_1 maps E(I) continuously into R equipped with this topology:

<u>(4.13) Remark:</u> $A_1 : E(I) \longrightarrow (R,\tau)$ cont.

5. BOUNDARY OPERATORS

Let $y,z \in R$. Then, by definition of R, there exists $w \in E(I)$ with $F_1 y = G_1 w$. In view of (3.2) z lies in E(I). Therefore, lemma 3.4 implies the integrability of $(S_1 z)^* F_1 y$. Furthermore, it can be shown that

$$(5.1)\quad [F_1 y, S_1 z]_{[\alpha,\beta]} - (y,z)_{[\alpha,\beta]} =$$

$$= z^*(\beta)\begin{pmatrix} 0 & E_k \\ 0 & 0 \end{pmatrix} y(\beta) - z^*(\alpha)\begin{pmatrix} 0 & E_k \\ 0 & 0 \end{pmatrix} y(\alpha).$$

Since $(S_1 z)^* F_1 y$ is integrable and since y and z ly in E(I) by (3.2), the left side of (5.1) converges for $\alpha \longrightarrow a$, $\beta \longrightarrow b$. This proves the existence of the limit

$$(5.2)\quad \langle y,z\rangle := -\lim_{\substack{\alpha\to a\\ \beta\to b}}\{z^*(\beta)\begin{pmatrix} 0 & E_k \\ 0 & 0 \end{pmatrix} y(\beta) - z^*(\alpha)\begin{pmatrix} 0 & E_k \\ 0 & 0 \end{pmatrix} y(\alpha)\}$$

for $y,z \in R$. Taking adjoints in (5.2) we see that the following limit exists for $y,z \in R$, too:

$$(5.3)\quad \{y,z\} := \lim_{\substack{\alpha\to a\\ \beta\to b}}\{z^*(\beta)\begin{pmatrix} 0 & 0 \\ E_k & 0 \end{pmatrix} y(\beta) - z^*(\alpha)\begin{pmatrix} 0 & 0 \\ E_k & 0 \end{pmatrix} y(\alpha)\} = -\overline{\langle z,y\rangle}.$$

Finally, for $\alpha\to a$ and $\beta\to b$ we obtain from (5.1):

$$(5.4)\quad (y,z)_I = [F_1 y, S_1 z]_I + \langle y,z\rangle \qquad (y,z \in R).$$

Now let $r(\leq 2k)$ be the dimension of $E_o(I)$. Since $(\ ,\ )_I$ is positive-definite on $E_o(I)$, there exists an orthonormal basis $y_1,\dots,y_r$ of $E_o(I)$. Denote $B := (y_1,\dots,y_r)$. Then the Gram-matrix of B equals E_r:

$$(5.5)\quad (B,B)_I = ((y_j,y_i)_I) = E_r.$$

Define $\langle y,B\rangle$, $\{y,B\}$ and $(y,B)_I$ to be the column-vectors with components $\langle y,y_i\rangle$, $\{y,y_i\}$ and $(y,y_i)_I$, resp. Then for $y \in R$ (5.3), (5.4) and $F_1 y_i = 0$ imply

$$\{y,B\} = (\{y,y_i\}) = -(\overline{\langle y_i,y\rangle}) = (-\overline{(y_i,y)_I}) = -(y,B)_I.$$

Especially, $\{y,B\}$ may be continuously continued to the closure $\overline{R}$ of R in $(E(I), \|\ \|_I)$:

$$\{y,B\} = -(y,B)_I \qquad (y \in \overline{R}).$$

To define boundary operators we choose (r,r)-matrices P,Q, V_o,V_1 and L which fulfill the following

(5.6) Assumptions: 1) P and Q are orthogonal projections with $P + Q = E_r$,
2) V_o and V_1 are hermitian and $V_oQ=V_1Q=0$, $V_o \geq 0$,
3) L is regular.

Then we may define boundary operators F_2,S_2 and G_2 as follows:

(5.7) Definition:

$$\left.\begin{aligned} F_2y &:= (Q - V_o)L\{y,B\} + PL^{*-1}\langle y,B\rangle \\ S_2y &:= -PL\{y,B\} + QL^{*-1}\langle y,B\rangle \end{aligned}\right\} \text{ for } y \in R$$

$$G_2y := -V_1L\{y,B\} \quad \text{for } y \in \bar{R}.$$

Indeed, by definition of $\langle y,B\rangle$ and $\{y,B\}$, these operators mapping R (resp. $\bar{R}$) into $\mathbb{C}^r$ depend only on the values of y near the boundary of I.

6. THE BOUNDARY-VALUE PROBLEM

We now define

$$Fy := \begin{pmatrix} F_1y \\ F_2y \end{pmatrix}, \quad Sy := \begin{pmatrix} S_1y \\ S_2y \end{pmatrix} \text{ for } y \in R, \quad Gy := \begin{pmatrix} G_1y \\ G_2y \end{pmatrix} \text{ for } y \in \bar{R}.$$

Then F and S are linear mappings from R into $\mathcal{L} \times \mathbb{C}^r$, G maps $\bar{R}$ linearly into $\mathcal{L} \times \mathbb{C}^r$.

We shall investigate the spectral theory of the boundary-value problem

(6.1) $Fy = \lambda Gy$.

If $G_2 \neq 0$, the boundary conditions in (6.1) depend on λ; if $G_2 = 0$ they reduce to the λ-independent boundary conditions $F_2y = 0$.

Now define

$\hat{R} := \{y \in \bar{R} \mid QL\{y,B\} = 0\}$.

The boundary conditions defining $\hat{R}$ are generalizations of the "essential" boundary conditions known from variational problems and from regular left-definite eigenvalue-problems.

It can be shown that

(6.2) $\{z,B\}^* L^* V_o L\{y,B\} = (S_2z)^* F_2y - \langle y,z\rangle$

if y and z belong to $R \cap \hat{R}$. Adding (5.4) to (6.2) we get

(6.3) $(y,z)_I + \{z,B\}^* L^* V_o L\{y,B\} = [F_1y, S_1z]_I + (S_2z)^* F_2y.$

The left side of (6.3) is defined for all $y,z \in \hat{R}$, and since V_o is positive-semidefinite by (5.6),2), it is an inner product on $\hat{R}$ denoted by (y,z). $\hat{R}$ can be shown to be complete with respect to this inner product. Furthermore, the positive-definiteness of V_o implies

(6.4) $(y,y)_I \leq (y,y).$

If we define

$$[f,g] := [f_1,g_1]_I + g_2^* f_2 \text{ for } f = \begin{pmatrix} f_1 \\ f_2 \end{pmatrix},\ g = \begin{pmatrix} g_1 \\ g_2 \end{pmatrix} \in \mathfrak{B} \times \mathbb{C}^r$$

the right side of (6.3) becomes $[Fy,Sz]$, so that (6.3) may be written as

(6.5) $[Fy,Sz] = (y,z) \quad (y,z \in R \cap \hat{R}).$

(6.5) justifies the notation "left-definite", since the inner product is built up by the left side of (6.1).

Define

$$F_2B := (F_2y_1,\dots,F_2y_r),\ S_2B := (S_2y_1,\dots,S_2y_r).$$

Then (5.5),(6.4),(6.5) and $F_1y_i = 0$ imply

$$E_r = (B,B)_I \leq (B,B) = (S_2B)^* F_2B.$$

Therefore, F_2B is regular and we may define

$$Ay := A_1y + B(F_2B)^{-1}(G_2y - F_2A_1y) \quad (y \in \hat{R}).$$

We have the following

(6.6) Theorem: A is a linear, continuous and selfadjoint map of $\hat{R}$ into $R \cap \hat{R}$. If $R \cap \hat{R}$ is equipped with the topology τ of local uniform convergence, A is continuous with respect to this topology. For each $y \in \hat{R}$

(6.7) $FAy = Gy$.

We shall partially prove this theorem: By theorem 4.12

$F_1 Ay = F_1 A_1 y = G_1 y$ and

$F_2 Ay = F_2 A_1 y + G_2 y - F_2 A_1 y = G_2 y$.

Therefore, Ay is a solution of the inhomogeneous boundary value problem (6.7). A obviously is linear. We remark that $[Gy, Sz] = [Sy, Gz]$ and thus (6.5) and (6.7) give

$$(Ay, z) = [FAy, Sz] = [Gy, Sz] = [Sy, Gz] =$$
$$= \overline{[Gz, Sy]} = \overline{[FAz, Sy]} = \overline{(Az, y)} = (y, Az).$$

Since A is defined on all of $\hat{R}$ this shows that A is selfadjoint and continuous.

7. SPECTRAL THEORY

In view of (6.7) and since by (6.5), (6.4) and the definiteness of $(,)_I$ on $E_o(I)$ F is injective on $R \cap \hat{R}$, the boundary value problem (6.1) is completely described by A. Therefore, the spectral theory of (6.1) is equivalent to that of A. We restrict A to the subspace $\overline{A\hat{R}}$ of $\hat{R}$. Then A remains bounded and selfadjoint. Furthermore, we define

$\mathfrak{N}: = \{ y \in \hat{R} \mid (y, y) = 0 \}$

and denote the canonical homomorphism of $\hat{R}$ onto $\hat{R}/\mathfrak{N}$ by φ. Then $\hat{R}/\mathfrak{N}$ is a Hilbert-space which contains $\varphi(\overline{A\hat{R}})$ as a sub-Hilbert-space. Then it is possible to show that

$\varphi(A(\overline{A\hat{R}})) = \varphi(A\hat{R})$

which therefore is dense in $\varphi(\overline{A\hat{R}})$, and that there exists a selfadjoint operator Γ in $\varphi(\overline{A\hat{R}})$ with domain $\varphi(A(\overline{A\hat{R}}))$ such that the following diagram commutes:

$$\begin{array}{ccccccc}
\hat{R} & \supset & \overline{A\hat{R}} & \xrightarrow{\;A\;} & A(\overline{A\hat{R}}) & \subset & \overline{A\hat{R}} \\
\downarrow \varphi & & \downarrow \varphi & & \downarrow \varphi & & \downarrow \varphi \\
\hat{R}/\mathfrak{N} & \supset & \varphi(\overline{A\hat{R}}) & \xleftarrow[\;\Gamma\;]{} & \varphi(A(\overline{A\hat{R}})) = \varphi(A\hat{R}) & \subset & \varphi(\overline{A\hat{R}})
\end{array}$$

Denote the resulution of the identity of Γ by E. Then E can be calculated. The result is the following

<u>(7.1) Theorem:</u> There exists a monotonously increasing map $\mathfrak{P}$ from $\mathbb{R}$ into the set of hermitian (2k,2k)-matrices with the following properties:

1) $\mathfrak{P}$ is continuous from the right, $\mathfrak{P}(0) = 0$

2) Let Y_λ denote the fundamental matrix of the initial value problem

$$F_1 Y_\lambda = \lambda G_1 Y_\lambda, \quad Y_\lambda(x_o) = E_{2k}.$$

Then $\int_o^\sigma Y_\tau d\mathfrak{P}(\tau)$ exists in the norm-topology of $\hat{R}^{2k}$ and lies in $(A(\overline{A\hat{R}}))^{2k}$. For $y \in \overline{A\hat{R}}$ and $\mu, \lambda \in \mathbb{R}$

$$\int_\mu^\lambda Y_\sigma d_\sigma (y, \int_o^\sigma Y_\tau d\mathfrak{P}(\tau))$$

exists in the norm-topology of $\hat{R}^{2k}$ and lies in $A(\overline{A\hat{R}})$.

3) For $y \in \overline{A\hat{R}}$ and $\mu, \lambda \in \mathbb{R}$ we have

(7.2) $$(E(\lambda) - E(\mu))\varphi(y) = \varphi(\int_\mu^\lambda Y_\sigma d_\sigma (y, \int_o^\sigma Y_\tau d\mathfrak{P}(\tau))).$$

Now for $\lambda \to \infty$, $\mu \to -\infty$, the left side of (7.2) tends to $\varphi(y)$. This implies the

<u>(7.3) Norm-expansion theorem:</u> For $y \in \overline{A\hat{R}}$

$$||y - \int_{-\infty}^{\infty} Y_\sigma d_\sigma (y, \int_o^\sigma Y_\tau d\mathfrak{P}(\tau))|| = 0.$$

The integral converges in the norm-topology of $\hat{R}$ and is an element of $\overline{A\hat{R}}$.

To obtain a direct expansion theorem we remark that by (6.4), (3.4), (4.11) and by the definition of A $Ay = 0$ for

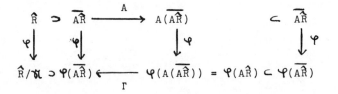

$y \in \mathfrak{M}$. Therefore there exists a mapping T from $\varphi(\overline{A\hat{R}})$ into $A(\overline{A\hat{R}})$ such that the diagram

$$\begin{array}{ccc} \overline{A\hat{R}} & \xrightarrow{\quad A \quad} & A(\overline{A\hat{R}}) \\ \varphi \big\downarrow & \overset{T}{\nearrow} & \big\downarrow \varphi \\ \varphi(\overline{A\hat{R}}) & \xleftarrow[\quad \Gamma \quad]{} & \varphi(A(\overline{A\hat{R}})) \end{array}$$

commutes, too. For $y \in A(\overline{A\hat{R}})$ this gives

(7.4) $$y = T\Gamma\varphi(y) = T\int_{-\infty}^{\infty} \sigma dE(\sigma)\varphi(y).$$

In view of theorem 6.6 A - and therefore T - is continuous if we consider on $A(\overline{A\hat{R}})$ the topology τ of local uniform convergence. Since the integral in (7.4) converges in the norm-topology of $\varphi(\overline{A\hat{R}})$,

$$\lim_{\substack{\mu \to -\infty \\ \lambda \to \infty}} T\int_{\mu}^{\lambda} \sigma dE(\sigma)\varphi(y)$$

exists in the topology τ. Finally it can be proved that

$$T\int_{\mu}^{\lambda} \sigma dE(\sigma)\varphi(y) = \int_{\mu}^{\lambda} Y_\sigma d_\sigma(y, \int_{0}^{\sigma} Y_\tau d\mathfrak{P}(\tau)).$$

This yields the

<u>(7.5) Direct-expansion theorem:</u> For $y \in A(\overline{A\hat{R}})$ and $x \in I$

$$y(x) = \int_{-\infty}^{\infty} Y_\sigma(x) d_\sigma(y, \int_{0}^{\sigma} Y_\tau d\mathfrak{P}(\tau))$$

and the integral converges locally uniformly on I.

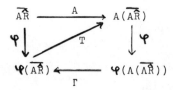

REFERENCES

1 NIESSEN, H.D., Singuläre S-hermitesche Rand-Eigenwertprobleme, manuscripta math. 3, 35 - 68 (1970)

2 NIESSEN, H.D., Zum verallgemeinerten zweiten Weyl'schen Satz, Arch.d.Math., 22, 648-656 (1971)

3 SCHÄFKE, F.W., and A. SCHNEIDER, S-hermitesche Rand-Eigenwertprobleme I, Math. Ann. 162, 9-26 (1965)

4 SCHÄFKE, F.W., and A. SCHNEIDER, S-hermitesche Rand-Eigenwertprobleme II, Math. Ann. 165, 236-260 (1966)

5 SCHÄFKE, F.W., and A. SCHNEIDER, S-hermitesche Rand-Eigenwertprobleme III, Math. Ann. 177, 67-94 (1968)

6 SCHNEIDER, A. and H.D. NIESSEN, Linksdefinite singuläre kanonische Eigenwertprobleme I. To appear in Journ.f.d. reine und angew. Math.

7 SHOTWELL, D.A., Boundary problems for the differential equation $Lu = \lambda\sigma u$ and associated eigenfunction-expansions, Univ. of Colorado, Thesis 1965

8 WEYL, H., Über gewöhnliche Differentialgleichungen mit Singularitäten und die zugehörigen Entwicklungen willkürlicher Funktionen, Math.Ann. 68, 220-269 (1910)

9 WEYL, H., Über gewöhnliche lineare Differentialgleichungen mit singulären Stellen und ihre Eigenfunktionen. Nachr. v.d.kgl. Gesellschaft der Wissenschaften zu Göttingen, Math.Physik.Klasse, Heft 5, 442-467 (1910).

FB Math.-Naturw.
Gesamthochschule
43 Essen
Kopstadtplatz 13
Germany

FB Math.-Naturw.
Gesamthochschule
56 Wuppertal 1
Hofkamp 82-86
Germany

SPECTRAL THEORY FOR LEFT-DEFINITE SINGULAR SYSTEMS OF DIFFERENTIAL EQUATIONS II.

by H.D. NIESSEN and A. SCHNEIDER

ABSTRACT: An extension of the direct-expansion theorem with regard to the "first" components will be proved. Then we discuss the integraltransformations and their inversion formulas, associated with the considered boundary-value problems.

1. INTRODUCTION.

In this second part of our paper we consider canonical systems

(1.1) $y_1' - D_{12}^{*}y_1 - D_{22}y_2 = 0$

(1.2) $y_2' + D_{11}y_1 + D_{12}y_2 = \lambda E_{11}y_1$

on an arbitrary interval I and assume, that all the assumptions, given in section 2. of the first part, are fulfilled. For these systems we had defined left-definite boundary-value problems and the spectral theory of these problems was reduced to the study of a selfadjoint operator in a suitable Hilbert-space. The main result is the representation of the resolution of the identity $E(\lambda)$ in terms of a fundamental matrix Y_λ of (1.1),(1.2):

(1.3) $$(E(\beta)-E(\alpha))\varphi(y) = \varphi\left(\int_\alpha^\beta Y_\sigma d\left(y,\int_0^\sigma Y_\tau d\mathfrak{P}(\tau)\right)\right);\quad y \in \overline{A\hat{R}}.$$

A consequence of (1.3) is the expansion theorem

(1.4) $$\left|\left|y-\int_{-\infty}^{-\infty} Y_\sigma d\left(y,\int_0^\sigma Y_\tau d\mathfrak{P}(\tau)\right)\right|\right| = 0 \qquad ;\ y \in A\hat{R}$$

which also implies the direct-expansion theorem

(1.5) $$y(x) = \int_{-\infty}^{+\infty} Y_\sigma(x) d\left(y,\int_0^\sigma Y_\tau d\mathfrak{P}(\tau)\right) \qquad ;\ y \in A(\overline{A\hat{R}}),$$

where the integral converges locally uniformly on I.

In this note we start by showing that (1.5) can be extended to the space $\overline{A\hat{R}}$ with regard to the "first" components of y. The proof follows from an inequality between the corresponding seminorms. The application to the special case of an ordinary second order differential equation yields the known result of Weyl [4].

Then we study the integraltransformations and their inversion formulas associated with these boundary-value problems in the case $V_o = 0$. Using matrix notation we get the analogue results of [1]. We conclude by pointing out, how the results of Shotwell [3] can be summarized in our theory.

Complete proofs of all results are given in [2], which will be published elsewhere.

2. A FUNDAMENTAL INEQUALITY.

Let $\mathfrak{Y}$ be the subspace of E(I) defined by

$$\mathfrak{Y}:=\{u|u\in E(I): u_1 \text{loc.abs.cont.} u_1'-D_{12}^{*}u_1=D_{22}u_2 \text{a.e.}\}.$$

We denote by $|u_1(x)|$ the Euclidian norm of $u_1(x)$. Then we can prove

<u>(2.1) Theorem:</u> For every compact subinterval $J\subset I$ there exists a $\delta(J) \geq 0$ such that for all $u\in\mathfrak{Y}$ the inequality

$$(2.2) \quad |u_1|_J:=\max_{x\in J}|u_1(x)| \leq \delta(J)||u||_I$$

is valid.

We don't give the details of the proof, but will point out the principle, on which the proof is based.

1. If

$N:=\{u|u\in\mathfrak{A}: F_1u=0\}$,

then there exists a subinterval $J_o\subset I$ such that $(u,v)_J$ is a positive definite hermitian scalarproduct on N for all intervals J with $J_o\subset J\subset I$. This is a simple consequence of our assumption (5) of section 2 of the first part.

2. Let $u\in\mathfrak{Y}$ and $J=[\alpha,\beta]\subset I$. If U(x) is the fundamental matrix of $Z'-D_{12}^{*}Z = 0$ with $U(\alpha) = E_k$, we define

$$(2.3) \quad v_1(x):=U(x)\int_{\alpha}^{x}U^{-1}(t)D_{22}(t)u_2(t)dt.$$

Then v_1 is locally absolutely continuous, $v_1'-D_{12}^{*}v_1=D_{22}u_2$ a.e. and with some $\alpha(J) \geq 0$

$$(2.4) \quad |v_1|_J \leq \alpha(J)||u||_J$$

is valid. Defining $v:=\begin{pmatrix} v_1 \\ u_2 \end{pmatrix}$, we have

(2.5) $||v||_J \leqq \tau(J)||u||_J$.

3. Let

(2.6) $w_1:=u_1-v_1$; $w:=\begin{pmatrix} w_1 \\ 0 \end{pmatrix}$.

Then

$$w_1 \in M:=\{Z|Z \text{ cont.diff}, Z'-D_{12}^{*}Z = 0\}.$$

For $J \supset J_0$ $|Z|_J$ and $(\int_J Z^{*}(x)D_{11}(x)Z(x)dx)^{1/2}$ are norms on M. Since the dimension of M is finite, these norms are equivalent on M and so we get with a constant $\beta(J) \geq 0$ the estimation

(2.7) $|w_1|_J \leqq \beta(J)(\int_J w_1^{*}(x)D_{11}(x)w_1(x)dx)^{1/2} = \beta(J)||w||_J$.

Now $u = v+w$ and the proof is complete.

3. THE DIRECT-EXPANSION THEOREM.

With the fundamental matrix $U(x)$ of the previous section we can represent each element $v \in \mathfrak{V}$ by

(3.1) $v_1(x) = U(x)v_1(\alpha)+U(x)\int_\alpha^x U^{-1}(t)D_{22}(t)v_2(t)dt$.

Hence we have together with the inequality (2.2)

<u>(3.2) Lemma:</u> For every u, which lies in the closure of $\mathfrak{V}$ with respect to the norm $||\ ||_I$, there exists an element $v \in \mathfrak{V}$ with $||v-u||_I = 0$. If $w \in \mathfrak{V}$ and $||w-u||_I = 0$ too, then $w_1 = v_1$.

If $\mathfrak{Z}$ is the vectorspace of all mappings v_1 from I into $\mathbb{C}^k$ such that with some mapping v_2 from I into $\mathbb{C}^k$ $v:=\begin{pmatrix} v_1 \\ v_2 \end{pmatrix}$ belongs to $\mathfrak{V}$, the lemma (3.2) permits to define a mapping P from $\overline{\mathfrak{V}}$ on $\mathfrak{Z}$ in the following manner. For $u \in \overline{\mathfrak{V}}$ we choose $v \in \mathfrak{V}$ with $||u-v||_I = 0$ and set $Pu:=v_1$. P is a linear mapping from $\overline{\mathfrak{V}}$ on $\mathfrak{Z}$ with $Pv=v_1$ for $v \in \mathfrak{V}$. P is continuous if $\mathfrak{Z}$ is equipped with the topology of locally uniform convergence. This follows from

<u>(3.3) Lemma:</u> For $u \in \overline{\mathfrak{V}}$ we have

(3.4) $|Pu|_J \leqq \delta(J)||u||_I$.

In the subspace $\bar{\bar{R}}$ the two norms $\|\ \|_I$ and $\|\ \|$ are equivalent and further $A\hat{R} \subset \hat{R} \subset \bar{\bar{R}}$. Hence the closure of $A\hat{R}$ with respect to the norm $\|\ \|$ is a subspace of $\hat{R}$. On $\hat{R}$ $\|u\|_I \leqq \|u\|$ and thus we get

(3.5) Lemma: For each compact subinterval $J \subset I$ there exists a $\delta(J) \geqq 0$ such that for all $u \in \overline{A\hat{R}}$

(3.6) $|Pu|_J \leqq \delta(J)\|u\|$.

Now we get the extension of the expansion theorem by

(3.7) Theorem: For $u \in \overline{A\hat{R}}$ we have the expansion theorem

(3.8) $$(Pu)(x) = \int_{-\infty}^{+\infty} (Y_\sigma)_1(x)\,d(u, \int_0^\sigma Y_\tau\, d\mathfrak{P}(\tau)).$$

$(Y_\lambda)_1(x)$ is the (k,2k)-matrix consisting of the first k rows of $Y_\lambda(x)$. The integral converges locally uniformly on I.

4. WEYL'S "POLAR" EQUATION.

Let q and k be real-valued functions on $I=[0,\infty)$ with

(4.1) $0 \not\equiv |k(x)| \leqq q(x)$

and p a positive, continuously differentiable function. Then for the special canonical system

(4.2) $y_1' + \frac{1}{p}y_2 = 0$

(4.3) $y_2' + qy_1 = \lambda k y_1$

the basic assumptions are given with

$$D_{11}(x) := q(x) \ ; \quad D_{22}(x) := -\frac{1}{p(x)}$$

$$D_{12}(x) := 0 \ ; \quad E_{11}(x) := k(x).$$

Obviously the system (4.2),(4.3) is equivalent to

(4.4) $y_2 = -(py_1')$

(4.5) $-(py_1')' + qy_1 = \lambda k y_1$.

Now $\dim E_0(I_0) = 2$ and we assume that $\dim E_0(I_\infty) = 1$. Weyl has proved in [4] , that the differential equation (4.5) for $\lambda = 0$ has a nontrivial (real-valued) solution β with

(4.6) $\lim_{x\to\infty} p(x)\beta(x)\beta'(x) = 0.$

We set

$$b(x) := \begin{pmatrix} \beta(x) \\ -p(x)\beta'(x) \end{pmatrix}$$

and get $F_1 b = 0$ and

(4.7) $||b||_I^2 = \int_0^\infty (p(x)(\beta'(x))^2+q(x)(\beta(x))^2)dx = -p(0)\beta'(0)\beta(0).$

We also may assume, that $||b||_I^2 = 1$. Hence $\{b\}$ is a normed base in $E_o(I)$.

Now we define boundary-mappings corresponding to section 5 of the first part by choosing the projection $Q = 1$, so that $P = V_o = V_1 = 0$. Since $\beta'(0) \neq 0$ by (4.7) we can set $L = (p(0)\beta'(0))^{-1}$. Then the boundary-mappings

$F_2 y := L\{y,b\}$; $y \in R$

$S_2 y := L^{-1}\langle y,b\rangle$; $y \in R$

$G_2 y := 0$; $y \in \overline{R}$

are of the form of definition (5.7) of the first part of our note. For $y \in R$ we have

$\lim_{x\to\infty} p(x)\beta'(x)y_1(x) = 0,$

so that $F_2 y = y_1(0)$ $(y \in R)$.

We denote by D the space of functions on I

$D := \{u | u \text{ diff}; u' \text{loc.abs.cont.}; \int_0^\infty (p|u'|^2+q|u|^2)dx<\infty\}.$

Then we have the

(4.8) Theorem: Let u,v and w be complex-valued functions on I with

(4.9) $\sqrt{q}\cdot w \in L^2(I)$,

(4.10) $v \in D$, $v(0)=0$ and $-(pv')'+qv=kw$ a.e.

(4.11) $u \in D$; $u(0) = 0$ and $-(pu')'+qu=kv$ a.e.

If $\eta_\lambda(x)$ and $\zeta_\lambda(x)$ are the solutions of (4.5) with

$\eta_\lambda(0) = 1 \quad ; \quad \eta'_\lambda(0) = 0$

$\zeta_\lambda(0) = 0 \quad ; \quad -p(0)\zeta'_\lambda(0) = 1$

and $Y_\lambda(x)$ is the matrix

$$Y_\lambda(x) := \begin{pmatrix} \eta_\lambda(x) & , \zeta_\lambda(x) \\ -p(x)\eta'_\lambda(x) & ,-p(x)\zeta'_\lambda(x) \end{pmatrix}$$

then we have in terms of the spectral distribution matrix $\mathfrak{P}(\tau)$, belonging to the boundary-value problem $Fy=\lambda Gy (y \in R)$, the locally uniformly convergent expansion

$$(4.12) \quad u(x) = \int_{-\infty}^{+\infty} (\eta_\sigma(x), \zeta_\sigma(x)) d\left(\begin{pmatrix} u \\ -pu' \end{pmatrix}, \int_0^\sigma Y_\tau d\mathfrak{P}(\tau) \right).$$

The proof follows from the fact, that by our assumptions we have

$$\begin{pmatrix} u \\ -pu' \end{pmatrix} \in \overline{A\hat{R}}.$$

If we assume, that the zeros of k don't have a finite accumulationpoint, then for every function m the function $\frac{1}{k} \cdot m$ is defined a.e. We abbreviate $-(pu')'+qu$ by Lu and hence the assumptions can be written in the (not quite exact) way:

1.) $u \in D \quad ; u(0) = 0$

2.) $\frac{1}{k}Lu \in D \quad ; (Lu)(0) = 0$

3.) $q^{1/2} \cdot \frac{1}{k}L(\frac{1}{k}Lu) \in L^2(I)$.

5. INTEGRALTRANSFORMATIONS

If f(t), g(t) and h(t) are matrix-valued functions on some bounded interval $J:=[\alpha,\beta]$, we introduce an integral of Riemann-Stieltjes-type by

$$(5.1) \quad \int_J f(t)(dg(t))h(t) := \lim_{\mu(\mathfrak{Z})\to 0} \sum_r f(\tau_r)(g(t_r)-g(t_{r-1}))h(\tau_r),$$

where the t_ρ are the points of the partition $\mathfrak{Z}$ and the τ_ρ are such that $t_{\rho-1} \leq \tau_\rho \leq t_\rho$. $\mu(\mathfrak{Z})$ is the mesh of $\mathfrak{Z}$. We also write $\int_\alpha^\beta f(t)(dg(t))h(t)$ instead of $\int_J f(t)(dg(t))h(t)$. Naturally the corresponding number of rows and columns must be equal. The integral exists, if f and h are continuous and g is of bounded variation on J. If f or h is the identity matrix, we simply write $\int_J f(t)dg(t)$ or $\int_J (dg(t))h(t)$ respectively. All the known theorems for Stieltjes-integrals can be proved. So we have the formula

$$(5.2)\quad \int_J f(t)(dg(t))h(t) = \int_J f(t)d(\int_\gamma^t (dg(s))h(s))$$

The matrix $\mathfrak{P}(t)$ is of bounded variation on every compact interval J of the reals and thus we are able to define for continuous f and h the hermitian sesquilinear form.

$$(5.3)\quad \mathfrak{P}_J(f,g) := \int_J g^*(t)d\mathfrak{P}(t)f(t).$$

Obviously $\mathfrak{P}_J(f,f) \geq 0$ and for $J_1 \subset J_2$ we get $\mathfrak{P}_{J_1}(f,f) \leq \mathfrak{P}_{J_2}(f,f)$. Let

$$(5.4)\quad C^2_{\mathfrak{P}} := \{f \mid f: \mathbb{R} \to \mathbb{C}^{2k},\ f \text{ cont},\ \mathfrak{P}_J(f,f) \text{ bounded}\}.$$

For f and $g \in C^2_{\mathfrak{P}}$ the limit

$$(5.5)\quad \mathfrak{P}(f,g) := \lim_{J\to\mathbb{R}} \mathfrak{P}_J(f,g)$$

exists by Schwarz's inequality and (5.5) defines a positive semidefinite hermitian scalarproduct on $C^2_{\mathfrak{P}}$. Now we complete $C^2_{\mathfrak{P}}$, form the quotientspace with respect to the subspace of all elements of vanishing norm and we receive the Hilbertspace $L^2_{\mathfrak{P}}$. We identify the equivalence-classes with the functions, denote the scalarproduct again by $\mathfrak{P}(f,g)$ and write for $f,g \in L^2_{\mathfrak{P}}$

$$\int_{\mathbb{R}} g^*(t)d\mathfrak{P}(t)f(t) := \mathfrak{P}(f,g)$$

Since the matrix $Y_\lambda(x)$ depends continuously on (λ,x), we get the

<u>(5.6) Lemma:</u> If $g \in E(I)$ and J is a compact subinterval of I, then the mapping f_J defined by

$$f_J(\mu):=(g,Y_\mu)_J = (\chi_J g,Y_\mu)_I$$

is continuous. Here χ_J is the characteristic function of J.

(5.7) Lemma: If $g\in\overline{A\hat{R}}$ and J is a compact subinterval of I, then we have

(5.8) $\chi_J g\in E(I)$.

$\chi_J g$ can be decomposed into

(5.9) $\chi_J g = v_J + w_J$

with $v_J\in \overline{A\hat{R}}$ and $w_J\in E(I)\ominus\overline{A\hat{R}}$, and for $J\to I$ v_J tends to g in the norms $\|\ \|_I$ and $\|\ \|$.

We now assume, that $V_o = 0$. Hence $(u,v) = (u,v)_I$ on $\overline{R}$ and we get

(5.10) Theorem: Let $g\in\overline{A\hat{R}}$ and J a compact subinterval of I. Then for f_J, defined by $f_J(\mu):=(\chi_J g,Y_\mu)$ we have

1.) $f_J\in C^2_{\mathfrak{P}}$,

2.) $\{f_J \mid J\subset I, J \text{ comp}\}$ is a Cauchynet for $J\to I$.

3.) If we define $Tg:=\lim_{J\to I} f_J$, then T is a linear isometric mapping from $\overline{A\hat{R}}$ into $L^2_{\mathfrak{P}}$.

We give a short sketch of the proof. The continuity of f_J follows from lemma (5.6). If $\chi_J g = v_J + w_J$ and $J_1 = [\gamma,\delta]$ is a compact subinterval of $\mathbb{R}$, we get by a simple computation

$$(5.11)\quad \mathfrak{P}_{J_1}(f_J,f_J) = \Big(\int_{J_1} Y_\mu d\big(v_J,\int_0^\mu Y_\sigma d\mathfrak{P}(\sigma)\big),\chi_J g\Big)$$
$$= ((E(\delta) - E(\gamma))\varphi(v_J),\varphi(v_J))$$
$$= \|(E(\delta) - E(\gamma))\varphi(v_J)\|^2.$$

This implies $f_J\in C^2_{\mathfrak{P}}$ and

(5.12) $\mathfrak{P}(f_J,f_J) \leq \|\chi_J g\|^2$.

For $J\subset J'\subset I$ we deduce

(5.13) $\mathfrak{P}(f_{J'}-f_J,f_{J'}-f_J) \leq \|\chi_{J'-J}g\|^2 \longrightarrow 0 \ (J\to I)$

and together with (5.11) we see, that

(5.14) $\mathfrak{P}(Tg,Tg) = \lim_{J\to I}\mathfrak{P}(f_J,f_J) = \lim_{J\to I}||v_J||^2 = ||g||^2.$

Dividing the fundamental matrix $Y_\lambda(x)$ up into

(5.15) $Y_\lambda(x) = \begin{pmatrix} (Y_\lambda)_1(x) \\ (Y_\lambda)_2(x) \end{pmatrix}$

we write

(5.16) $(Tg)(\mu):=\int_I((Y_\mu)^*_1(x)D_{11}(x)g_1(x)-(Y_\mu)^*_2(x)D_{22}(x)g_2(x))dx,$

where the integral converges in the norm-topology of $L^2_{\mathfrak{P}}$.

6. THE INVERSION FORMULAS

<u>(6.1) Theorem:</u> If $h \in L^2_{\mathfrak{P}}$ and J_1 is a compact interval, we set

(6.2) $h_{J_1}(x):=\int_{J_1} Y_\mu(x)(d\mathfrak{P}(\mu))h(\mu).$

Then we have

(6.3) $h_{J_1} \in \overline{A\hat{R}}$.

(6.4) $\{h_{J_1}|J_1 \subset \mathbb{R}, J_1$ compact$\}$ is a Cauchynet for $J_1 \to \mathbb{R}$ and the mapping V, defined by

(6.5) $Vh:=\lim_{J_1\to\mathbb{R}} h_{J_1}$

is a linear and isometric mapping from $L^2_{\mathfrak{P}}$ into $\overline{A\hat{R}}$.

(6.5) is also written in the form

$(Vh)(x) =\int_{\mathbb{R}} Y_\mu(x)d\mathfrak{P}(\mu)h(\mu),$

and the convergence of the integral is understood in the norm-topology of $\overline{A\hat{R}}$.

The mappings T and V are reciprocal in the sense of

<u>(6.6) Theorem:</u> For $g \in \overline{A\hat{R}}$ we have

(6.7) $\varphi(VTg) = \varphi(g)$.

By theorem (6.6) we can prove, that the mapping T is surjective. The injectivity of T is not given in general. For V we can deduce the injectivity from

(6.8) Theorem: The mapping T and V satisfy

(6.9) $TV = id_{L^2_{\mathfrak{p}}}$.

If we know, that

$$\mathfrak{N} := \{y \mid y \in \hat{R} : \|y\| = 0\} = \{0\},$$

then φ is injective on $\overline{A\hat{R}}$ and hence

(6.10) $VT = id_{\overline{A\hat{R}}}$.

Thus T and V are bijective and are reciprocal to each other. In this case (6.9) and (6.10) will be expressed by the formulas

(6.11) $f(\mu) := \int_I ((Y_\mu)^*_1(x) D_{11}(x) g_1(x) - (Y_\mu)^*_2(x) D_{22}(x) g_2(x))dx$

(6.12) $g(x) = \int_{\mathbb{R}} Y_\mu(x) d\mathfrak{p}(\mu) f(\mu)$.

7. THE PROBLEM OF SHOTWELL.

In [3] D.A. Shotwell has studied boundary value problems associated with the ordinary 2k-order differential equation

(7.1) $\sum_{\kappa=0}^{k} (-1)^\kappa (p_\kappa \eta^{(\kappa)})^{(\kappa)} = \lambda q \eta.$

Equation (7.1) can be written as a canonical system by setting

$$D_{11}(x) := \begin{pmatrix} p_{k-1}(x) & & 0 \\ & \ddots & \\ 0 & & p_0(x) \end{pmatrix}$$

$$E_{11}(x) := \begin{pmatrix} 0 & \ddots & & 0 \\ & & 0 & \\ 0 & & & q(x) \end{pmatrix}$$

$$D_{22}(x) := \begin{pmatrix} -p_k(x)^{-1} & & 0 \\ & 0 & \\ & & \ddots \\ 0 & & & 0 \end{pmatrix}$$

$$D_{12}(x) := \begin{pmatrix} 0 & 1 & & & 0 \\ & \ddots & \ddots & & \\ & & & \ddots & 1 \\ 0 & & & & 0 \end{pmatrix}$$

and the basic assumptions are given, if we make the following assumptions.

1.) The functions p_κ are real-valued, κ-times continuously differentiable functions with

$p_\kappa(x) \geqq 0$; $(\kappa=0,1\ldots k-1)$

$p_k(x) > 0$

2.) q is real-valued, continuous and

$0 \not\equiv |q(x)| \leqq p_o(x)$

Then the formulas (6.11)(6.12) yield the corresponding results concerning the Parseval's relation, and the norm-expansion theorem is given under much weaker conditions than made by Shotwell. In contrary to Shotwell we don't need any assumption of differentiability of q in some subinterval or any hypothesis on the defiency index.

REFERENCES

[1] NIESSEN, H.D. and A. SCHNEIDER, Integraltransformationen zu singulären S-hermiteschen Rand-Eigenwertproblemen, manuscripta math. 5, 133-145 (1971)

[2] SCHNEIDER, A. and H.D. NIESSEN, Linksdefinite singuläre kanonische Eigenwertprobleme II, in preparation.

[3] SHOTWELL, D.A., Boundary problems for the differential equation $Lu = \lambda\sigma u$ and associated eigenfunction-expansions, Univ. of Colorado, Thesis (1965)

[4] WEYL, H., Über gewöhnliche lineare Differentialgleichungen mit singulären Stellen und ihre Eigenfunktionen, Nachr. v.d.kgl. Gesellschaft der Wissenschaften zu Göttingen, Math. Physik Klasse, Heft 5, 442-467 (1910)

FB Math.-Naturw.
Gesamthochschule
43 Essen
Kopstadtplatz 13
Germany

FB Math.-Naturw.
Gesamthochschule
56 Wuppertal
Hofkamp 82-86
Germany

THE LIMIT-POINT CLASSIFICATION OF DIFFERENTIAL EXPRESSIONS

J. B. McLEOD
Mathematical Institute
Oxford

1. Introduction

We shall be concerned with properties of the differential operator $M[\cdot]$ with

$$M[f] = -(pf')' + qf \quad \text{on} \quad [0,\infty) \left(' \equiv \frac{d}{dx}\right), \tag{1.1}$$

where the coefficients p and q are real-valued on $[0,\infty)$ and satisfy the following conditions:

(i) p is absolutely continuous and strictly positive,

(ii) q is locally integrable.

Much of what we have to say is applicable also to the case where the operator is defined on an interval $[0,b)$, where b is now finite but is in some sense a singularity of the coefficients p,q; we do not however follow up the details of this, which can be found, along with a more detailed account of other points in this note, in a forthcoming paper [1] by W.N. Everitt, M. Giertz and the present author.

In 1910 Weyl [2] showed that if λ is any given non-real number, then the equation

$$M[y] = \lambda y \tag{1.2}$$

has at least one non-trivial (i.e. not identically zero) solution which lies in $L^2(0,\infty)$, and this leads to a classification of such differential operators. The operator M is said to be <u>limit-point</u> (LP) if there is just one linearly independent solution of (1.2) in $L^2(0,\infty)$, and to be <u>limit-circle</u> (LC) if two linearly independent solutions (and so all) lie in $L^2(0,\infty)$. Which case arises is independent of the choice of λ, so long as λ is non-real, and in the limit-circle case, all solutions of (1.2) are $L^2(0,\infty)$ even when λ is allowed to be real; in the limit-point case, the only assertion that can be made for real λ is that <u>at most</u> one linearly independent solution is $L^2(0,\infty)$. The terms limit-point and limit-circle arise because of the elegant mode of proof employed by Weyl, which relates the problem to a sequence of circles, each of which lies inside the previous one so that the sequence converges either to a limit-point or to a limit circle.

This classification of M depends only on the nature of the coefficients p and q. There are no known necessary and sufficient conditions on p and q to distinguish between LP and LC but there is a necessary and sufficient condition in terms of certain functions in $L^2(0,\infty)$. Define the linear manifold $D(p,q) \subset L^2(0,\infty)$ to consist of those functions $f \in L^2(0,\infty)$ for which f' is absolutely continuous and $M[f] \in L^2(0,\infty)$. By integration by parts, we have, for any $f,g \in D(p,q)$,

$$\int_0^X \{\bar{g}\, M[f] - f\, M[\bar{g}]\}dx = [p(f\bar{g}' - f'\bar{g})]_0^X, \tag{1.3}$$

so that

$$\lim_{\infty} p(f\bar{g}' - f'\bar{g}) \qquad (1.4)$$

exists and is finite for all $f,g \in D(p,q)$. It can further be shown that M is LP if and only if the limit (1.4) is zero for all $f,g \in D(p,q)$ (see, for example, [3] or section 18.3 of [4].)

These ideas can be linked to the terminology for linear operators in $L^2(0,\infty)$. Thus, if we take as the domain D_M of the operator M those functions in $D(p,q)$ which have the additional property that

$$f(0) \cos \alpha + f'(0) \sin \alpha = 0, \qquad (1.5)$$

then the integrated term at the lower limit drops out in (1.3) if $f,g \in D_M$, and the limit in (1.4) being zero for such f,g therefore implies that the operator M, with domain D_M, is symmetric. It can further be proved that it is in fact self-adjoint, and so, in the LP case, we have a self-adjoint operator, and so a spectral theory, naturally associated with the operator M and the boundary condition (1.5). In the LC case this natural association breaks down, and to associate a specific self-adjoint operator with M, it is necessary to impose, in addition to (1.5), a boundary condition at ∞.

Because of the implications of the LP-LC classification for spectral theory, much work has gone into the investigation of conditions on p and q which will guarantee one case or the other. Some examples of results of this sort follow below, although it must be emphasised that they are not an exhaustive list and are chosen rather because of their relevance for the remainder of the paper.

Theorem A. *If p and q satisfy the conditions in (i) and (ii) above, and if q is bounded below, then M is LP.* (This result was known to Weyl.)

Proof. The result is almost trivial if we consider the equation

$$M[y] = \lambda y \qquad (1.6)$$

where λ is chosen sufficiently large and negative that $q(x) > \lambda$ for $x \in [0,\infty)$. Then, for the solution of (1.6) for which $y(0) = y'(0) = 1$, we know that $(py')'$ always has the same sign as y, and so py' never vanishes and y is steadily increasing and so certainly not in $L^2(0,\infty)$. Thus M is LP.

Theorem B. *If $p=1$, q satisfies the conditions in (ii) above, and $q(x) \geqslant -Q(x)$, where Q is positive non-decreasing and satisfies*

$$\int^{\infty} Q^{-\frac{1}{2}} dx = \infty,$$

then M is LP. In particular, if $p=1$ and $q(x) \geqslant -x^2$, then M is LP. (This result was first proved by Titchmarsh (see section 12.11 of [5]); a more general result (for partial differential operators) is due to Ikebe and Kato [6].)

Theorem C. If $p=1$, q satisfies the conditions in (ii) above, and q is bounded below in a sequence of non-overlapping intervals $[x_{2n}, x_{2n+1}]$, where the sequence $\{x_n\}$ is increasing and $x_{2n+1} - x_{2n} \to 0$ as $n\to\infty$, then M is LP. (This result is due to Hartman [7], and has since been generalised by Eastham [8]; it is possible to give an extension for partial differential operators, and this will be done in a forthcoming paper by Eastham, W.D. Evans, and the present author.)

Theorem D. Suppose $p=1$ and $q(x) \to -\infty$ as $x \to \infty$, with

$$q'(x) < 0 \quad \text{and} \quad q'(x) = O\{|q(x)|^c\} \qquad (0 < c < \tfrac{3}{2}),$$

and let $q''(x)$ be ultimately of one sign. Then if

$$\int^\infty |q|^{-\frac{1}{2}}dx < \infty,$$

M is LC.

This result is due to Titchmarsh [9]. It is proved by finding asymptotic expansions for solutions of the equation $M[y] = \lambda y$ as $x\to\infty$, this being possible because of the smoothness conditions on q. In fact, we find that there are solutions y with

$$y(x) \sim \{\lambda-q(x)\}^{-\frac{1}{4}} \exp\{\pm i\xi(x)\},$$

where

$$\xi(x) = \int_0^x\{\lambda-q(t)\}^{\frac{1}{2}}dt.$$

Since the LP case corresponds to the condition that (1.4) be zero for all $f,g \in D(p,q)$, the question arises whether there is any significance in the simpler condition that

$$\lim_\infty pf\bar{g}' = 0 \tag{1.7}$$

for all $f,g \in D(p,q)$. (It will be clear that satisfaction of (1.7) necessarily implies that (1.4) is zero, although the contrary is not *a priori* necessarily the case.) This leads to the following definition: M is strongly limit-point (SLP) if (1.7) is satisfied for all $f,g \in D(p,q)$; M is weakly limit-point (WLP) if (1.4) is zero for all $f,g \in D(p,q)$ but (1.7) is not satisfied, so that for some pair $f,g \in D(p,q)$ the limit in (1.7) either does not exist or, if it does exist, is not zero.

The purpose of this note is to discuss conditions under which M is SLP or WLP, and where possible to relate these to the results in Theorems A-D. To summarise the results very roughly, we can see (as has already been pointed out) that SLP imples LP, and that also, at least under several 'reasonable' conditions, the opposite is true (cf. Theorems 1 and 3 below); at the same time it is possible to construct examples which are only WLP as in §§5-7, and also counter-examples to other related (and superficially reasonable) conjectures which can be made in this same area.

2. A lemma

We start with an almost trivial lemma which recurs frequently in the remainder of the paper.

Lemma 1. For the general differential operator M of §1, if it is known that M is LP, then it is sufficient, in order to establish that M is SLP, to show that

$$\lim_{\infty} pFF' = 0$$

for all real $F \in D(p,q)$.

Proof. Suppose that

$$\lim_{\infty} pFF' = 0$$

for all real $F \in D(p,q)$, and take any real pair $f,g \in D(p,q)$; put $F = f+g$, and then

$$pFF' = p(ff' + gg' + f'g + fg'),$$

and so

$$p(f'g + fg') \to 0 \quad \text{as} \quad x\to\infty.$$

But M is LP, and so

$$p(f'g = fg') \to 0 \quad \text{as} \quad x\to\infty,$$

from which $pf'g \to 0$ and (1.7) is satisfied for real functions. By writing any complex function in $D(p,q)$ in the form $f_1 + if_2$, f_1 and f_2 real, we can then show that (1.7) is satisfied for complex-valued functions and so that M is SLP. This proves the sufficiency of the condition, and the necessity is trivial.

3. The case q bounded below

Theorem 1. Let M be given by (1.1), where p and q satisfy the conditions (i) and (ii) of §1; if additionally q is bounded below, then M is SLP.

Proof. It is no loss of generality to suppose $q \geqslant 0$. Then, by integration by parts, and restricting ourselves to real f in accordance with Lemma 1, we have

$$\int_0^X fM[f]dx = \int_0^X \{qf^2 + pf'^2\}dx - [pff']_0^X, \tag{3.1}$$

so that, if $f \in D(p,q)$, the left hand side converges as $X\to\infty$, and so also therefore does the right-hand side. Hence either

$$\begin{array}{l} \text{(i) } p^{\frac{1}{2}}f',\ q^{\frac{1}{2}}f \in L^2(0,\infty), \text{ and} \\ \text{(ii) } \lim_{\infty} pff' \text{ exists and is finite} \end{array} \tag{3.2}$$

or the integral on the right of (3.1) diverges to $+\infty$ as $X\to\infty$ and

$$\lim_{\infty} pff' = +\infty.$$

But this second alternative is impossible, since it implies that ultimately f and f' are of the same sign, and then $f \notin L^2(0,\infty)$; and we can conclude therefore that (3.2) holds. Further, the limit in (ii) of (3.2) must be negative or zero, since the assumption that it is positive would again imply that $f \notin L^2(0,\infty)$.

Suppose then for contradiction that the limit is negative. Then ultimately we have, say, $f > 0$, $f' < 0$, pf' non-decreasing. The first two inequalities imply that f tends to a finite limit, necessarily zero if $f \in L^2(0,\infty)$, and the second two unequalities imply that pf' tends to a finite limit. Hence $pff' \to 0$, as required.

4. The case q unbounded below

If q is unbounded below, it is no longer nessarily true that M is SLP, and a specific counter-example is given in §5. But it is interesting to consider what results corresponding to Theorem 1 hold when we suppose, for example, not that q is bounded below, but that

$$q(x) \geqslant -kx^2 \tag{4.1}$$

for some positive constant k, a condition which by Theorem B guarantees that the problem is LP. What we obtain is

Theorem 2. Let M be given by (1.1) with p=1 and q satisfying (ii) of §1; if additionally q satisfies (4.1), then for all $f \in D(1,q)$,

$$\text{(i)}\quad x^{-1}f' \in L^2(1,\infty),\quad x^{-1}|q|^{\frac{1}{2}}f \in L^2(1,\infty), \tag{4.2}$$

$$\text{(ii)}\quad \lim_{\infty} x^{-2}ff' = 0. \tag{4.3}$$

Proof. Let $f \in D(1,q)$ and be real-valued; then from the identity

$$f'^2 + (q+kx^2)f^2 = (ff')' + M[f]f + kx^2f^2,$$

we obtain, on dividing by x^2 and integrating over $[1,X]$,

$$\int_1^X\{x^{-2}f'^2 + (x^{-2}q+k)f^2\}dx = \int_1^X x^{-2}(ff')'dx + \int_1^X x^{-2}fM[f]dx + $$
$$+ k\int_1^X f^2dx$$
$$= [x^{-2}ff']_1^X + 2\int_1^X x^{-3}ff'dx + \int_1^X x^{-2}fM[f]dx + k\int_1^X f^2dx$$
$$\leqslant X^{-2}f(X)f'(X) + O(\{\int_1^X x^{-2}f'^2dx\}^{\frac{1}{2}}) + O(1)$$

as $X\to\infty$. The integrand on the left is non-negative, and it is easy to see that if this integral does not converge, then $f(X)\,f'(X)$ is strictly positive for all large X, which is inconsistent with $f \in L^2(0,\infty)$. Arguing now with the real and imaginary parts of an arbitrary $f \in D(1,q)$, we obtain (4.2), and also that

$$\lim_{\infty} x^{-2} ff^{1}$$

exists and is finite. We can then argue as in Theorem 1 that this limit must be zero.

This result is best possible, and to this we return in §7.

At the same time, if we restrict q by (4.1), and at the same time impose on it the smoothness conditions of Theorem D, the resulting differential expression is SLP. Precisely we have

Theorem 3. Let M be given by (1.1) with $p=1$ and q satisfying the conditions of Theorem D except that now

$$\int^{\infty} |q|^{-\frac{1}{2}} dx = \infty.$$

Then M is SLP.

Proof. Let $F \in D(1,q)$, and let $\lambda = \mu + i\nu$ be any complex number with $\nu > 0$; let $\phi(\cdot,\lambda)$ and $\psi(\cdot,\lambda)$ be solutions of the differential equation

$$M[y] = -y + q(x)y = \lambda y \quad \text{on} \quad [0,\infty)$$

which satisfy

(i) $\phi(0,\lambda) = 1, \quad \phi'(0,\lambda) = 0;$

(ii) $\psi(\cdot,\lambda) \in L^2(0,\infty);$

(iii) $\phi(x,\lambda)\,\psi'(x,\lambda) - \phi'(x,\lambda)\,\psi(x,\lambda) = 1 \quad (x \in [0,\infty)).$

The existence of such a solution $\psi(\cdot,\lambda)$ follows from the analysis given in Chapter II of [9].

Now, for any function $f \in L^2(0,\infty)$, define

$$\Phi(x,\lambda,f) = \psi(x,\lambda) \int_0^x \phi(t,\lambda)\, f(t)dt + \phi(x,\lambda) \int_x^{\infty} \psi(t,\lambda)\, f(t)dt.$$

If both f and $M[f]$ are in $L^2(0,\infty)$, then Lemma 2.9 of [9] assures us that

$$\Phi(x,\lambda,f) = \frac{1}{\lambda}\{f(x) + \Phi(x,\lambda, M[f])\},$$

so that

$$f(x) = \Phi(x,\lambda, \lambda f - M[f]).$$

Taking $f=F$, and setting $\Psi = \lambda F - M[F]$, we thus have

$$F(x) = \psi(x,\lambda) \int_0^x \phi(t,\lambda)\Psi(t)dt + \phi(x,\lambda) \int_x^{\infty} \psi(t,\lambda)\Psi(t)dt$$

and by differentiation

$$F'(x) = \psi'(x,\lambda) \int_0^x \phi(t,\lambda)\Psi(t)dt + \phi'(x,\lambda) \int_x^{\infty} \psi(t,\lambda)\Psi(t)dt.$$

For fixed λ and large x the asymptotic nature of the solutions ϕ and ψ may be determined from section 22.2b of [4] or from section 5.10 of [9]; if for notational simplicity we consider just the case $q(x) = -x^2$, then (recalling that $\nu = \text{im}\lambda > 0$) we have

$$\phi(x,\lambda) = O(x^{-\frac{1}{2}+\frac{1}{2}\nu}), \quad \phi'(x,\lambda) = O(x^{\frac{1}{2}+\frac{1}{2}\nu}),$$

$$\psi(x,\lambda) = O(x^{-\frac{1}{2}-\frac{1}{2}\nu}), \quad \psi'(x,\lambda) = O(x^{\frac{1}{2}-\frac{1}{2}\nu}),$$

as $x\to\infty$. Thus in order to prove

$$\lim_{\infty} FF' = 0,$$

it is sufficient to show that

$$\lim_{\infty} x^{\frac{1}{2}\nu}\int_x^{\infty} t^{-\frac{1}{2}-\frac{1}{2}\nu}\,|\Psi(t)|dt = 0$$

and

$$\lim_{\infty} x^{-\frac{1}{2}\nu}\int_1^{x} t^{-\frac{1}{2}+\frac{1}{2}\nu}\,|\Psi(t)|dt = 0.$$

The first result follows from the Cauchy-Schwarz inequality, remembering that $\Psi \in L^2(0,\infty)$. To deal with the second, we divide the integral into integrals over [1,A] and [A,x]. Then, given $\varepsilon > 0$, we can make the second term in modulus not exceeding $\frac{1}{2}\varepsilon$ by the Cauchy-Schwarz inequality and sufficiently large choice of A, and then the first term in modulus not exceeding $\frac{1}{2}\varepsilon$ by sufficiently large choice of x.

It is also worth remark that what appears to be important in the choice of q in order to give the SLP case is not that q be monotonic, but that, if it oscillates, it should do so smoothly and not too rapidly. Thus if for example $q(x) = h(x)\sin x$, where h(x) is a suitably monotonic function, then the asymptotic formulae given in [10] suffice to show that again the resulting operator M is SLP. We omit the details.

5. An example that is WLP

In the differential operator M of §1, take $p(x)=1$, and for notational convenience consider the interval $[1,\infty)$ instead of $[0,\infty)$. Consider the function f defined by

$$f(x) = \frac{2 + \cos(x^3)}{x},$$

and define $\quad q(x) = f''(x)/f(x)$.

With p and q so determined, the resulting differential operator M is LP; this follows since one solution of $M[y] = 0$ is $y_1 = f$, with f given above, whilst a second linearly independent solution y_2 is given by

$$y_2(x) = f(x)\int_1^x f^{-2}(t)dt,$$

and a simple calculation shows that $y_2 \notin L^2(1,\infty)$.

Further $f \in L^2(1,\infty)$, $M[f]=0$, and so $f \in D(1,q)$. But

$$\lim_{\infty} ff' \neq 0$$

(in fact the limit does not exist), and so M is not SLP.

6. An example where q satisfies (4.1) and M is WLP.

The example of §5 leaves some questions unanswered. Thus the function q given there is highly oscillatory, since $q(x)$ contains a term involving $x^4 \cos(x^3)$. Since $q(x) = -x^2$ is something of a marginal case in the area, and in view of Theorem 3, there arises the question whether the WLP case can arise with q satisfying (4.1). We now construct an example to show that indeed it can.

As in §5, we take $p(x)=1$; q is defined by

$$q(x) = \begin{cases} n^2, & x \in [x_{2n}, x_{2n+1}], \\ -\nu_n^2, & x \in (x_{2n+1}, x_{2n+2}), \end{cases}$$

where $\{\nu_n : n = 0,1,2,...\}$ is a non-negative sequence, bounded away from zero and to be specified in detail later. (We shall however find that ν_n is of order n) The sequence $\{x_r : r = 0,1,2,...\}$ is to have the properties that

$$x_0 = 0, \quad x_{2n+1} - x_{2n} = 1 \quad (n = 0,1,2,...),$$

while

$$x_{2n+2} - x_{2n+1} = 1 + O(n^{-1}) \quad \text{as} \quad n \to \infty$$

with the precise nature of the O-term again to be specified later.

In the interval $[x_{2n}, x_{2n+1}]$ any real solution of the differential equation $M[y] = y'' + qy = 0$ has the form (A_n and B_n real numbers)

$$y(x) = A_n e^{nx} + B_n e^{-nx}$$

and in (x_{2n+1}, x_{2n+2}) the form (C_n and E_n real numbers)

$$y(x) = C_n \cos(\nu_n x + E_n).$$

The successive values for A_n, B_n, C_n, E_n are required to be so chosen that y and y' are continuous at the points $\{x_n : n = 1,2,3,...\}$. In order to produce the required example it will be sufficient, as in §5, to produce a solution y with the properties that

(i) $y \in L^2(0,\infty)$,

(ii) $\nu_n \leq Kn$ $(n = 0,1,2,...)$ for some positive K,

(iii) $yy' \not\to 0$ as $x \to \infty$.

We note that a necessary and sufficient condition for (i) to be satisfied is that

$$\sum_{n=0}^{\infty} C_n^2 < \infty. \tag{6.1}$$

For it is clear that this condition is equivalent to y being L^2 over the union of the intervals $[x_{2n+1}, x_{2n+2}]$; and in an interval $[x_{2n}, x_{2n+1}]$ the fact that y satisfies the equation $y'' - n^2 y = 0$ assures us that y cannot have a positive maximum (or a negative minimum) in (x_{2n}, x_{2n+1}), so that the maximum modulus of y must be attained at either x_{2n} or x_{2n+1}. The condition (6.1) then ensures that y is L^2 over the union of the intervals $[x_{2n}, x_{2n+1}]$, and the whole argument may be reversed.

We next observe that we may select the sequence $\{\nu_n\}$, with $\nu_n \asymp n$, so that the sequence $\{C_n\}$ satisfies

$$C_n = (n+1)^{-1} \text{ if } n \text{ is not a square;} \quad C(k^2) = (k+1)^{-1}.$$

(By $\nu_n \asymp n$ we mean that there exist constants K_1 and K_2 so that, for all sufficiently large n, $K_1 n \leq \nu_n \leq K_2 n$; and to avoid double suffixes, we write $C(k^2)$ for C_{k^2}.) For if we impose continuity at $x = x_{2n}$ and at $x = x_{2n+1}$, and eliminate A_n and B_n from the resulting equations, we obtain

$$e^n = C_n C_{n-1}^{-1}(\cos\theta_n - t_n \sin\theta_n)(\cos\phi_n - n^{-1}\nu_{n-1}\sin\phi_n)^{-1}, \quad (6.2)$$

$$e^{-n} = C_n C_{n-1}^{-1}(\cos\theta_n + t_n \sin\theta_n)(\cos\phi_n + n^{-1}\nu_{n-1}\sin\phi_n)^{-1}; \quad (6.3)$$

where

$$\theta_n = \nu_n x_{2n+1} + E_n, \quad t_n = n^{-1}\nu_n, \quad \phi_n = \nu_{n-1}x_{2n} + E_{n-1},$$

for all $n=1,2,\ldots$ To prove that we can make the required choice of $\{C_n\}$, we suppose inductively that the solution is determined up to the interval $[x_{2n-1}, x_{2n})$ with the required numbers $\{C_r: r=0,1,2,\ldots,n-1\}$. (Note that by starting the solution off suitably we can certainly arrange for the hypothesis to be satisfied for $n=1$.) The hypothesis implies of course that ν_{n-1}, E_{n-1} and also x_{2n-1} are all determined, but the value of x_{2n} is still at our choice. It will be sufficient to show that we can choose values of x_{2n}, ν_n and E_n so that (6.2) and (6.3) are satisfied with the specified values of C_{n-1} and C_n and the already determined values of ν_{n-1} and E_{n-1}.

To do this, we first choose x_{2n} $(= x_{2n-1} + 1 + O(n^{-1}))$ so that

$$(\cos\phi_n - n^{-1}\nu_{n-1}\sin\phi_n)e^n C_{n-1}C_n^{-1} = 1. \quad (6.4)$$

(Since $\nu_{n-1} \asymp n$, a variation of order n^{-1} in x_{2n} will allow any value whatever for $\tan\phi_n$, and so will enable (6.4) to be satisfied.) Then from (6.2)

$$\cos\theta_n - t_n \sin\theta_n = 1,$$

while from (6.3)

$$\cos\theta_n + t_n \sin\theta_n = o(1)$$

for large n. Hence for large n,

$$\cos\theta_n = \tfrac{1}{2} + o(1), \quad \sin\theta_n = -\tfrac{1}{2}\surd 3 + o(1),$$

$$t_n = {}^1/_{\surd 3} + o(1), \quad \nu_n = n\{{}^1/_{\surd 3} + o(1)\}.$$

We have thus found x_{2n}, ν_n and E_n to give the required value for C_n, and the process can be continued inductively.

The example is now virtually complete. For the solution y thus determined satisfies $y \in L^2(0,\infty)$ since $\Sigma C_n^2 < \infty$.

Further $yy' \not\to 0$ as $x \to \infty$; for in an interval of the form $[x(2k^2+1), x(2k^2+2)]$,

$$(yy')(x) = -\tfrac{1}{2}C^2(k^2)\nu(k^2) \sin\{2(\nu(k^2)x + E(k^2))\},$$

which does not tend to zero as $k \to \infty$ since

$$C(k^2) = (k+1)^{-1} \quad\text{and}\quad \nu(k^2) > K_1 k^2.$$

7. Examples to show that Theorem 2 is best possible

The technique adopted in §6 to obtain a counter-example can be extended to show that Theorem 2 is best possible, in the sense that, given any positive non-decreasing function $\chi(x)$ with $\chi(x)/x \to 0$ as $x \to \infty$, we can find a function q satisfying (4.1) and $f \in D(1,q)$ so that $\chi^{-1}f' \notin L^2(0,\infty)$; and similarly we can find a function q satisfying (4.1) and $f \in D(1,q)$ so that $\chi^{-2}ff' \not\to 0$ as $x \to \infty$. The examples are constructed using functions q which are step-functions, and the precise choice of q depends on the function χ. While the essential features are those of the example in §6, the details are rather more complicated, and will be found in [1].

REFERENCES

1. W.N. Everitt, M. Giertz and J.B. McLeod, "On the strong and weak limit-point classification of second-order differential expression", to be published by the London Math.Soc.
2. H.Weyl, "Über gewöhnliche Differential gleichungen mit Singularitäten und die zugehörigen Entwicklungen willkürlicher Funktionen", Math.Ann. 68(1910) 220-269.
3. W.N. Everitt, "A note on the self-adjoint domains of second order differential equations", Quart.J.Math. (Oxford) (2) 14(1963) 41-45.
4. M.A. Naimark, Linear differential operators : Part II (Ungar, New York, 1968).
5. E.C. Titchmarsh, Eigenfunction expansions associated with second-order differential equations : Part II (Oxford University Press, 1958).
6. T. Ikebe and T. Kato, "Uniqueness of the self-adjoint extension of singular elliptic differential operators, Arch.Rat.Mech.Anal. 9(1962) 77-92.
7. P. Hartman, "The number of L^2 - solutions of x"+q(t)x = 0", American J.Math. 73(1951) 635-645.
8. M.S.P. Eastham, "On a limit-point method of Hartman", Bull. London Math.Soc. 4(1972) 340-344.
9. E.C. Titchmarsh, Eigenfunction expansions associated with second-order differential equations : Part I (Oxford University Press, 1962).

10. J.B. McLeod, "On the spectrum of wildly oscillating functions", J.London Math.Soc. 39(1964) 623-634.

SECOND- AND FOURTH-ORDER DIFFERENTIAL EQUATIONS WITH OSCILLATORY COEFFICIENTS AND NOT OF LIMIT-POINT TYPE

M. S. P. EASTHAM
Mathematics Department, Chelsea College,
University of London

1. INTRODUCTION

We consider first the second-order self-adjoint differential expression $M = d^2/dx^2 + q(x) \quad (0 \leq x < \infty)$, (1.1)
where $q(x)$ is real-valued and locally Lebesgue integrable in $[0,\infty)$. The important limit-point, limit-circle classification of (1.1), due to Weyl, 1910, is as follows. Let λ be a complex parameter. Then M is said to be limit-circle if all solutions of the differential equation

$$My(x) = \lambda y(x) \tag{1.2}$$

are $L^2(0,\infty)$ and to be limit-point otherwise. This classification of M is independent of λ, and the fundamental theorem of Weyl is that, in the limit-point case, (1.2) has precisely one linearly independent $L^2(0,\infty)$ solution when λ is non-real. The terms limit-point and limit-circle arise from the use that Weyl made of a certain sequence of nested circles, the sequence converging in the limit to either a point or a circle. Accounts of the Weyl theory and its consequences in the spectral theory of (1.2) are given in several books, of which we refer here to Coddington and Levinson, 1955, Hille, 1969, and Titchmarsh, 1962.

No necessary and sufficient condition on $q(x)$ is known which distinguishes between the limit-point and limit-circle cases, but a number of sufficient conditions for the two cases are known. The first result which we give in this direction is as follows.

A. Let $q(x) < 0$ in some interval $[X,\infty)$ and let the product $(-q)^{-\frac{1}{4}}((-q)^{-\frac{1}{4}})''$ be $L(X,\infty)$. Then M is limit-circle or limit-point according as $(-q)^{-\frac{1}{4}}$ is $L^2(X,\infty)$ or is not $L^2(X,\infty)$.

This result is proved by taking $\lambda = 0$ in (1.2) and applying the Liouville transformation to obtain the asymptotic form of the solutions of $My(x) = 0$ - see Coppel, 1965 (p.120), Eastham, 1970 (p.61), and Wong, 1973. For a recent extension, see Knowles, 1973. A further result of the same nature but with conditions only on $q'(x)$ is in Atkinson, 1957.

The conditions on $q'(x)$ and $q''(x)$ implied in A can be thought of as restrictions on the oscillatory nature of $q(x)$ and therefore only a fairly restricted class of functions $q(x)$ is covered. An example is $q(x) = -x^{\alpha}$, which makes M limit-circle if $\alpha > 2$ and limit-point if $\alpha \leqslant 2$. However, as far as the investigation of the limit-point case is concerned, other methods are available and consequently, as we shall indicate in §2 below, a wide variety of functions $q(x)$ are known which make M limit-point. In contrast, the type of limit-circle in A was until recently the only one on the limit-circle case. This has, I think, created the impression that those $q(x)$ which behave badly in some oscillatory sense must necessarily make M limit-point. I hope to destroy this impression in this lecture.

2. LIMIT-POINT CONDITIONS

As I wish to concentrate on the limit-circle case, I give here only a brief survey of limit-point conditions. There are two well-known generalizations, B and C below, of the limit-point example $q(x) = -x^{\alpha}$ ($\alpha \leq 2$) already mentioned.

B. (Levinson, 1949, and Sears, 1950) M is limit-point if there is a continuous function $Q(x)$ (>0) such that $q(x) \geqslant -Q(x)$, $Q^{-\frac{1}{2}}(x)$ is not $L(0,\infty)$ and <u>either</u> $Q'(x) = o\{Q^{3/2}(x)\}$ as $x \to \infty$ <u>or</u> $Q(x)$ is non-decreasing.

C. (Hartman and Wintner, 1949) M is limit-point if

$$\int_0^X q_-(x)\, dx = O(X^3) \quad (X \to \infty),$$

where $$q_-(x) = \min\{q(x), 0\}.$$

It was Hartman, 1951, who first drew attention to a result of a different kind in which the conditions are imposed on $q(x)$ only in a sequence of intervals as follows.

D. (Hartman, 1951) M is limit-point if $q(x) \geqslant k$ (constant) only for x in a sequence of non-overlapping intervals of fixed length.

This covers certain oscillating function, e.g., $q(x) = x^{\alpha} \sin x$ where the intervals are $(2m\pi, 2(m+1)\pi)$ and $k = 0$. Further results of this kind have been developed by Ismagilov, 1962, Atkinson and Evans, 1972, and Eastham, 1972. Ismagilov considered the differential expression

$$M_n = (-1)^n d^{2n}/dx^{2n} + q(x)$$

of order 2n, for which we define the term limit-point in §5.

E. (Ismagilov, 1962) M_n is limit-point if there is a sequence of non-overlapping intervals (a_m, b_m) with $\sum (b_m - a_m)^{2n} = \infty$ and $q(x) \geqslant -k(b_m - a_m)^{-2n}$ in (a_m, b_m).

We shall refer to E again in §5, but we point out here that it does of course include D. Atkinson and Evans, 1972, showed that the Levinson-Sears conditions in B can be relaxed again to hold only in a suitable sequence of intervals.

The results B, D, and E give pointwise conditions on $q(x)$. An integral condition, like C, but again involving only a sequence of intervals is the following.

F. (Eastham, 1972) M is limit-point if there is a sequence of non-overlapping intervals (a_m, b_m) and a sequence of real numbers v_m such that $\sum v_m^{-1} = \infty$, $(b_m - a_m)^2 v_m \geqslant K > 0$, and

$$\int_{a_m}^{b_m} |q_-(x)| \, dx \leqslant k(b_m - a_m)^3 v_m^2.$$

This result F includes all the limit-point results for M referred to above, although the demonstration that it includes B is not quite straightforward and follows from work of Knowles, 197-. The additional results of Brinck, 1959, and Walter, 1968, are apparently not covered by F.

<u>An example</u> Let a, α, and β be constants with $\beta \leqslant 2$ and $|a| > 1$ Then

$$q(x) = x^{\alpha}\{-1 + a \sin(x^{\beta})\}$$

makes M limit-point for all α.

To show this, we use E. Let θ be the angle in $(-\frac{1}{2}\pi, \frac{1}{2}\pi)$ such that $\sin\theta = a^{-1}$ and define

$$a_m = (2m\pi + \theta)^{1/\beta}, \quad b_m = (2m\pi + \pi - \theta)^{1/\beta} \quad (a > 1),$$

$$a_m = (2m\pi - \pi - \theta)^{1/\beta}, \quad b_m = (2m\pi + \theta)^{1/\beta} \quad (a < -1).$$

Then $q(x) \geqslant 0$ in (a_m, b_m) and $b_m - a_m = O(m^{-1+1/\beta})$ as $m \to \infty$. Thus the conditions in E are satisfied (with $k = 0$) if $\beta \leqslant 2$.

Question Is M still limit-point if $|a| \leqslant 1$?

3. LIMIT-CIRCLE CONDITIONS

Moving on to limit-circle results, we begin by referring to Green's formula, valid for any $f(x)$ and $g(x)$ with locally absolutely continuous first derivatives in $[0,\infty)$:

$$\int_0^X \{\bar{g}(x)Mf(x) - f(x)M\bar{g}(x)\}\,dx = [f,g](X) - [f,g](0),$$

where $$[f,g] = f\bar{g}' - f'\bar{g}.$$

Hence, if $$f,\ g,\ Mf,\ Mg \text{ are all } L^2(0,\infty), \tag{3.1}$$

then $$[f,g](x) \to l,$$

a finite limit, as $x \to \infty$. In addition, as a result from general theory (Naimark, 1968, p.78), we have $l = 0$ for all f and g satisfying (3.1) if M is limit-point. Hence we have the following limit-circle condition:

(*) If $$[f,g](x) \to l \ (\neq 0) \text{ as } x \to \infty \tag{3.2}$$
for some f and g satisfying (3.1), then M is limit-circle.

This result (*) provides a method for investigating the limit-circle case which was used recently by Everitt, 1972, and Eastham, 1972 and 1973. This approach also has the advantage that it extends to the fourth-order case and this aspect will be described in §5 below. We use the method now to prove the following limit-circle result.

Theorem (Eastham, 1973) Let $P(x)$ and $h(x)$ be real-valued functions defined in an interval $[X,\infty)$ such that $P(x) > 0$, $P''(x)$ is continuous, and $P(x)$ and $h(x)$ are $L^2(X,\infty)$. In $[X,\infty)$, let $$q(x) = P^{-4}(x) + P''(x)/P(x) + h(x)/P(x) \tag{3.3}$$
and, in $[0,X]$, let $q(x)$ be $L^2(0,X)$. Then M is limit-circle.

In (3.2), we take

$$g(x) = f(x) = P(x)\exp\left(i\int_X^x Q(t)\,dt\right) \quad (x \geqslant X),$$

where $Q(x) = P^{-2}(x)$, while $f(x)$ and $g(x)$ are arbitrary in $[0,X)$. Then f is $L^2(0,\infty)$ and, in $[X,\infty)$,

$$|Mf| = |\{-P'' + (Q^2 + q)P - i(P^2Q)'/P\}| = |h|.$$

Hence Mf is $L^2(0,\infty)$ and so (3.1) holds. Also,

$$[f,g](x) = -2iP^2(x)Q(x) = -2i,$$

giving $l = -2i$ in (3.2). The theorem follows from (*).

Corollary 1. The choice $h(x) = 0$ gives the limit-circle example

$$q(x) = -P^{-4}(x) + P''(x)/P(x). \tag{3.4}$$

Corollary 2. Let $q(x) < 0$ in an interval $[X,\infty)$ and let both $(-q)^{-\frac{1}{4}}$ and $((-q)^{-\frac{1}{4}})''$ be $L^2(X,\infty)$. Then M is limit-circle.

In the Theorem, take $P(x) = (-q(x))^{-\frac{1}{4}}$ and $h(x) = -P''(x)$.

This corollary is virtually the limit-circle part of A in §1. The only difference is that the product $(-q)^{-\frac{1}{4}}((-q)^{-\frac{1}{4}})''$ is $L(X,\infty)$ in A, whereas here the separate factors are $L^2(X,\infty)$.

Corollary 3 (Eastham, 1973). Let $r(x)$, $s(x)$, and $H(x)$ be real-valued functions defined in an interval $[X,\infty)$ and such that $r(x) > 0$, $r''(x)$ and $s''(x)$ are continuous, $r(x)$ and $H(x)$ are $L^2(X,\infty)$, $s(x) \to 0$ as $x \to \infty$, and

$$r^{-3}(x)s(x),\ r'(x)s'(x),\ r(x)s(x)s''(x)$$

are all $L^2(X,\infty)$. Then

$$q(x) = -r^{-4}(x) + s''(x) + r^{-1}(x)H(x) \tag{3.5}$$

makes M limit-circle.

In the Theorem, we take $P = r(1 + s)$, where we can suppose that X is chosen large enough to make $s > -1$ in $[X,\infty)$, and

$$h = -r''(1 + s) - 2r's' + rss'' + r^{-3}(1 + s)^{-3}\{1 - (1 + s)^4\} + H(1 + s).$$

The conditions imposed on r and s ensure that h is $L^2(X,\infty)$. With our choice of h, (3.3) reduces to (3.5).

We note that, when $r(x) = x^{-\frac{1}{4}\alpha}$ $(\alpha > 2)$, (3.5) gives

$$q(x) = -x^{\alpha} + s''(x) + x^{\frac{1}{4}\alpha}H(x). \tag{3.6}$$

The conditions on $s(x)$ in Corollary 3 can be thought of as covering functions $s(x)$ which tend to zero as $x \to \infty$ while $s''(x)$ has large oscillations. Thus there is the possibility of obtaining rapidly oscillating limit-circle examples from (3.5) and this matter is taken up in the next section.

4. SOME APPLICATIONS

Application of Corollary 1 We prove the following result.

(Eastham and Thompson, 1973) Given any ϵ (> 0), there are functions $q_1(x)$ and $q_2(x)$ such that $q_1(x)$ makes M limit-point,

$q_2(x)$ makes M limit-circle, and

$$q_1(x) = q_2(x)$$

except in a sequence of intervals of total length at most ϵ. Further, $q_1(x)$ and $q_2(x)$ can be taken to be infinitely differentiable and $q_1(x)$ to be non-increasing.

We construct $q_1(x)$ and $q_2(x)$ from one and the same step-function $Q(x)$ by joining up the steps in two different ways. There is a whole class of functions $Q(x)$ which can be used in the construction but, to be specific, we take

$$Q(x) = -\pi^2 n^4 \quad (n - 1 \leq x < n;\ n = 1, 2, \ldots).$$

It is not difficult to check that $Q(x)$ makes M limit-point.

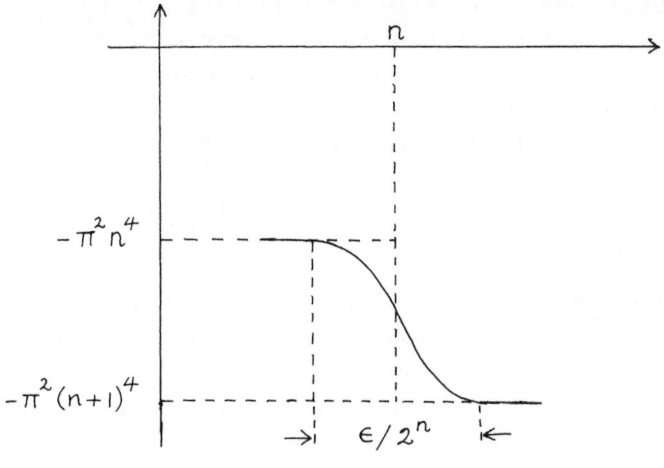

Fig. 1.

We now join up the steps as indicated in Fig. 1 to obtain a non-increasing infinitely differentiable $q_1(x)$ such that $Q(x) - q_1(x)$ is $L(0,\infty)$. It is then an easy consequence of the variation of constants formula that $q_1(x)$ also makes M limit-point. Now $(-q_1)^{-\frac{1}{4}}$ is $L^2(0,\infty)$ and we can therefore choose $P = (-q_1)^{-\frac{1}{4}}$ in (3.4) to define a limit-circle $q_2(x)$. As $P''(x) = 0$ on the straight parts of the graph of $q_1(x)$, (3.4) gives $q_1(x) = q_2(x)$ on the straight parts. This proves the result.

With our remarks at the end of §1 in mind, we note that, while it has been arranged that $q_1(x)$ is non-increasing, $q_2(x)$ has a large oscillation in the neighbourhood of $x = n$ (see

Fig. 2). This follows on observing the change of sign of $P''(x)$ in (3.4) near to $x = n$.

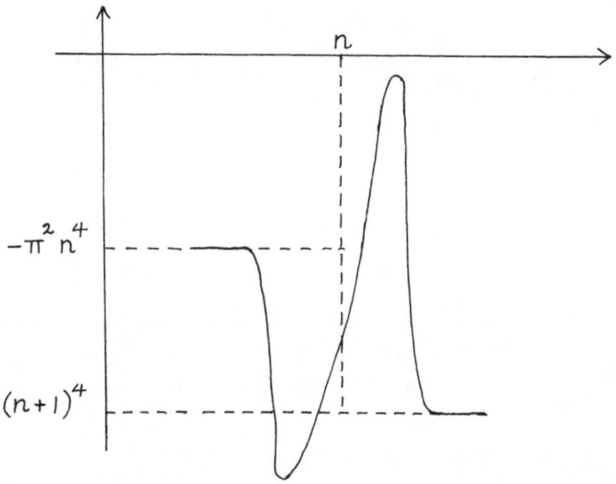

Fig. 2.

The result just proved makes it difficult to conceive what form the essential difference between the limit-point and limit-circle cases would take in terms of $q(x)$.

<u>Application of Corollary 3</u> In (3.6), we take

$$s(x) = -a\beta^{-2} x^{\alpha-2(\beta-1)} \sin(x^\beta),$$

where a is a constant and $\alpha > 2$. The conditions of Corollary 3 are satisfied if also

$$\beta > (7/8)\alpha + 5/4.$$

Then, with a suitable choice of $H(x)$, (3.6) gives

$$q(x) = x^\alpha\{-1 + a\sin(x^\beta)\}. \qquad (4.1)$$

This is the same $q(x)$ that was considered in the example at the end of §2, but there the conditions $\beta \leq 2$ and $|a| > 1$ were imposed. In Fig. 3 below, we illustrate the limit-point, limit-circle nature of (4.1) under the condition $|a| > 1$.

Again with the remarks at the end of §1 in mind, we note that the larger β is, the more rapidly does $q(x)$ oscillate. Thus, when $\alpha > 2$, it is rapid oscillations that give rise to the limit-circle case.

I have dwelt on this example here because it identifies a

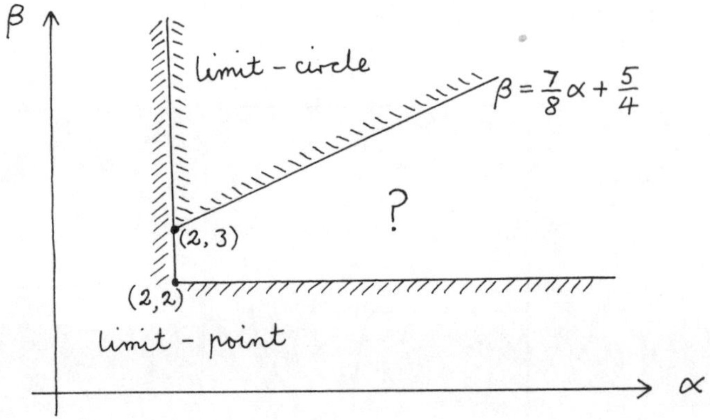

Fig. 3.

new type of behaviour which differentiates between the limit-point and limit-circle cases. The result A in 1 refers primarily to the <u>size</u> of $q(x)$. In the example, when $\alpha > 2$, the <u>frequency of oscillations</u> of $q(x)$ is the relevant factor - high frequency gives limit-circle, low frequency gives limit-point.

5. THE FOURTH-ORDER CASE

In order not to get involved with too many details, we consider here, not the most general fourth-order self-adjoint differential expression with real coefficients, but only the special case $M_1 = d^4/dx^4 + q(x) \quad (0 \leq x < \infty)$.
Much of what we say applies, or can be extended, to the general case. Let N denote the number of linearly independent $L^2(0,\infty)$ solutions of the fourth-order equation

$$M_1 y(x) = \lambda y(x) \quad (\text{im } \lambda \neq 0).$$

Then, from general theory, N is independent of λ and N = 2, 3, or 4 (Everitt, 1963, Naimark, 1968). We say that M_1 is <u>limit-N</u>; the term limit-point is also used for limit-2. Thus M_1 can be classified as limit-2, limit-3, or limit-4. (We note in parentheses that, for self-adjoint differential expressions of order 2n on $[0,\infty)$, the range for N is $n \leq N \leq 2n$. The case $N = n$ is called the limit-point case.)

If $q(x) = -x^{\alpha}$, asymptotic methods (Naimark, 1968, p.196) give

$$N = 2 \text{ if } \alpha \leqslant 4/3 \tag{5.1}$$

and

$$N = 3 \text{ if } \alpha > 4/3. \tag{5.2}$$

In recent years, several new methods have been developed for investigating the case N = 2 and we refer to Everitt, 197-, for a survey. We mention here only the earlier result of Ismagilov, 1962, that N = 2 if $q(x) \geqslant -x^{4/3}$. This generalizes (5.1) and is a consequence of E in §2 above.

There is much less information on the limit-3 case and we indicate now how the method in §3 above can be extended to M_1 to give a new class of functions q(x) which make N = 3. Full details will appear in Eastham, 197-.

The Green's formula for M_1 is

$$\int_0^X \{\bar{g}(x)M_1f(x) - f(x)M_1\bar{g}(x)\}\, dx = [f,g](X) - [f,g](0),$$

where

$$[f,g] = f'''\bar{g} - f''\bar{g}' + f'\bar{g}'' - f\bar{g}'''.$$

The formula holds for any f(x) and g(x) with locally absolutely continuous third-order derivatives. Corresponding to (*), we have

(**) If $$[f,g](x) \to 1 \ (\neq 0) \text{ as } x \to \infty \tag{5.3}$$
for some f and g such that f, g, M_1f, and M_1g are all $L^2(0,\infty)$, then M_1 is not limit-2.

The limit-3 case is obtained by combining (**) with

($\not\perp$) If $q(x) \leqslant 0$ in an interval $[X,\infty)$, then M_1 is not limit-4.

In (5.3), we take

$$g(x) = f(x) = P(x)\exp\left(i\int_0^x Q(t)\, dt\right),$$

where P and Q are real-valued and P > 0. Then

$$[f,f] = 2iU, \tag{5.4}$$

where $U = 2Q^3P^2 - 4QPP'' + 2QP'^2 - 2Q'PP' - Q''P^2$,

and $$|M_1f| = |\{(Q^4 + q)P - V + P^{(4)} - iU'/P\}|$$

where $$V = 6Q^2P'' + 2QQ'P' + 4QQ''P + 3Q'^2P.$$

For f and M_1f to be $L^2(0,\infty)$, we require that

$$P,\ U'/P,$$

and $$(Q^4 + q)P - V + P^{(4)} \quad (= h, \text{ say}) \tag{5.5}$$
are all $L^2(0,\infty)$. The rough idea is now this:

(i) choose an $L^2(0,\infty)$ function P;

(ii) choose $Q = P^{-2/3}$ so that (5.4) is $[f,f] = 4i + o(1)$ as $x \to \infty$ - conditions on P and its derivatives are implied here;

(iii) solve (5.5) for q to give
$$q = -Q^4 + (h + V - P^{(4)})/P. \tag{5.6}$$

As an example, the choice $P(x) = x^{-3\alpha/8}$ $(\alpha > 4/3)$, $h = P^{(4)} - V$ gives $q(x) = -x^{\alpha}$ - see (5.2) above - except that h is $L^2(0,\infty)$ only if $\alpha < 12$ (Eastham, 1971). However, a modification of the form $Q = P^{-2/3}(1 + T)$, where T is a certain o(1) function removes the condition $\alpha < 12$.

A result parallel to (3.6) is obtained by modifying P to $P = x^{-3\alpha/8}(1 + s)$, and $s^{(4)}(x)$ appears in place of $s''(x)$. The details are lengthy. Finally, we mention that a suitable choice of s(x) leads to the limit-3 example
$$q(x) = x^{\alpha}\{-1 + a\sin(x^{\beta})\},$$
where $\alpha > 4/3$, $\beta > (7/8)\alpha + 3/2$, and $a \leqslant 1$.

REFERENCES

Atkinson, F.V., 1957. Proc. Glasgow Math. Assoc. 3, 105.

Atkinson, F.V. and W.D.Evans, 1972. Math. Zeit. 127, 323.

Brinck, I., 1959. Math. Scand. 7, 219.

Coddington, E.A. and N.Levinson, 1955. Theory of ordinary differemtial equations (McGraw-Hill).

Coppel, W.A., 1965. Stability and asymptotic behaviour of differential equations (Heath).

Eastham, M.S.P., 1970. Theory of ordinary differential equations (Van Nostrand Reinhold).

---, 1971. J. London Math. Soc. (2) 3, 297.

---, 1972. Bull. London Math. Soc. 4, 340.

---, 1973. Quart. J. Math. (Oxford) (2) 24, 257.

---, 197-. J. London Math. Soc., to appear.

Eastham, M.S.P. and M.L.Thompson, 1973. Quart. J. Math. (Oxford) (2) 24, 531.

Everitt, W.N., 1963. Math. Annalen 149, 320.

---, 1972. Quart. J. Math. (Oxford) (2) 23, 193.

---, 197-. Niew Archief voor Wiskunde, to appear.

Hartman, P., 1951. American J. Math. 73, 635.

Hartman, P. and A.Wintner, 1949. American J. Math. 71, 206.

Hille, E., 1969. Lectures on ordinary differential equations (Addison-Wesley).

Ismagilov, R.S., 1962. Soviet Math. 3, 279.

Knowles, I., 1973. Quart. J. Math. (Oxford) (2) 24, 451.

---, 197-. J. London Math. Soc., to appear.

Levinson, N., 1949. Časopis pro Pěstováni Mat. Fys. 74, 17.

Naimark, M.A., 1968. Linear differential operators, Part 2 (Harrap).

Sears, D.B., 1950. Canadian J. Math. 2, 314.

Titchmarsh, E.C., 1962. Eigenfunction expansions, Part 1, 2nd edition (Oxford).

Walter, J., 1968. Math. Zeit. 105, 345.

Weyl, H., 1910. Math. Annalen 68, 222.

Wong, J.S.W., 1973. Quart. J. Math. (Oxford) (2) 24, 423.

SOME ASPECTS OF MULTI-PARAMETER SPECTRAL THEORY

B. D. SLEEMAN
Department of Mathematics,
University of Dundee, Dundee, Scotland

1. Introduction

In this paper we discuss some recent developments in the theory of multi-parameter eigenvalue or spectral problems. The typical situation is that in which we have a system of k (≥ 2) - second order ordinary differential operators depending on k-spectral parameters and defined on k-compact or non-compact intervals see equation (2.2) below. Such problems often arise in application of the separation of variables technique to boundary value problems for partial differential equations. Although the subject is historically almost as old as the now well established one-parameter case, it is only recently that it has aroused any real interest. For an introduction to the subject we refer to the review articles [1] and [10] and also to the recent book of Atkinson [2].

In section 2 of this paper we bring together and discuss some of the more important results obtained recently in what may be called the "regular" case, e.g. Sturm-Liouville problems. For example we discuss completeness and expansion theorems and to a lesser extent oscillation theory.

Sections 3 and 4 are devoted to the singular case in which we consider a generalisation of the well known Weyl limit-point limit-circle theory to the multiparameter case, while in the last section (section 5) we offer an alternative formulation of the problem.

2. Regular Problems

Let H_r, $r = 1,2,\dots,k$ be separable Hilbert spaces and H their tensor product. In each of the spaces H_r let T_r: $D(T_r) \subset H_r \to H_r$ be a self adjoint operator and let V_{rs}: $H_r \to H_r$, $s = 1,2,\dots,k$ be a collection of Hermitian operators. Also assume that each T_r has a compact resolvent and that zero is in the resolvent set of each T_r. Make also the hypothesis that

$$\det(V_{rs} f_r, f_r)_r > 0 \tag{2.1}$$

for all $f_r \neq 0$, $f_r \in H_r$. Here $(\, , \,)_r$ denotes the inner product in H_r.

We shall be concerned with the study of the system of simultaneous equations

$$T_r f_r + \sum_{s=1}^{k} \lambda_s V_{rs} f_r = 0 \, , \qquad r = 1,2,\ldots, k \tag{2.2}$$

in which $\lambda_1, \lambda_2, \ldots, \lambda_k$ are complex parameters. A k-tuple $(\lambda_1, \lambda_2, \ldots, \lambda_k)$ of complex numbers and a vector $f = f_1 \otimes f_2 \otimes \ldots \otimes f_k$ will be called respectively an eigenvalue and an eigenvector for the system (2.2) if $(\lambda_1, \lambda_2, \ldots, \lambda_k)$ and $f_1, f_2, \ldots, f_k$ satisfying (2.2) non-trivially.

The most complete result associated with (2.2) is due to Browne [3] and is contained in the following theorem

<u>Theorem 2.1</u>

With respect to the system (2.2) there is a set of decomposable eigenvectors

$$u_m = h_{m1} \otimes \ldots \otimes h_{mk} \, , \qquad m = 1,2,\ldots$$

with, for real k-tuples $(\lambda_{m1}, \ldots, \lambda_{mk})$,

$$T_r h_{mr} + \sum_{s=1}^{k} \lambda_{ms} V_{rs} h_{mr} = 0 \, , \qquad r = 1,2,\ldots,k$$

which are ortho-normal in the sense

$$\det(V_{rs} h_{mr}, h_{m'r})_r = \delta_{mm'} \, .$$

Here the eigenvalues are repeated according to multiplicity. For each

$V_{rs}: H_r \to H_r$, let V_{rs}^+ denote the linear map of H into H defined for decomposable tensors by

$$V_{rs}^+(f_1 \otimes \dots \otimes f_k) = f_1 \otimes \dots \otimes f_{r-1} \otimes V_{rs}\, f_r \otimes f_{r+1} \otimes \dots \otimes f_k$$

and extended to H by linearity and continuity. For each $T_r: D(T_r) \subset H_r \to H_r$, let T_r^+ denote the linear map of H into H defined on decomposable tensors $f = f_1 \otimes \dots \otimes f_k$ with $f_r \in D(T_r)$ by

$$T_r^+ f = f_1 \otimes \dots \otimes f_{r-1} \otimes T_r\, f_r \otimes f_{r+1} \otimes \dots \otimes f_k$$

and extended by linearity. Let $f = f_1 \otimes \dots \otimes f_k$ be such that each $f_r \in D(T_r)$ and such that there exist $g_1, g_2 \dots, g_k \in H$ satisfying the system

$$T_r^+ f + \sum_{s=1}^{k} V_{rs}^+ g_s = 0 , \qquad r = 1,2,\dots,k .$$

Then there holds the Parseval equality

$$\det(V_{rs}\, f_r, f_r)_r = \sum_m |\det(V_{rs}\, f_r, h_{mr})_r|^2 . \tag{2.3}$$

A result analogous to theorem 2.1 has also been established by Atkinson [2] under the hypothesis that the V_{rs} are compact relative to T_r .

Probably the most important special case of (2.2) is the multi-parameter Sturm-Liouville problem. Here we take $H_1 = H_2 = \dots = H_k = L^2(0,1)$ and let $T_r: D(T_r) \subset H_r \to H_r$ be the Sturm-Liouville operator

$$T_r \equiv \frac{d^2}{dx_r^2} - q_r(x_r) , \qquad 0 \leq x_r \leq 1 , \qquad r = 1,2,\dots,k$$

$$q_r(x_r) \in C[0,1] \quad \text{and real valued.}$$

whose domain

$$D(T_r) = \{y_r(x_r) \in L^2(0,1): \ (2.4) \text{ holding}\} ;$$

$$y_r(0)\cos\alpha_r - \frac{dy_r(0)}{dx_r}\sin\alpha_r = 0\ , \qquad 0 \leq \alpha_r < \pi\ ,$$
$$y_r(1)\cos\beta_r - \frac{dy_r(1)}{dx_r}\sin\beta_r = 0\ , \qquad 0 < \beta_r \leq \pi\ , \tag{2.4}$$

Thus each T_r is self-adjoint with compact resolvent. For the operator $V_{rs}: H_r \to H_r$ we take

$$(V_{rs}\, y_r)(x_r) = a_{rs}(x_r)y_r(x_r)\ ,$$

where

$$a_{rs}(x_r) \in C[0,1] \quad \text{and real valued}\ , \quad r,s = 1,\ldots,k\ .$$

Since a_{rs} is real valued and bounded V_{rs} is a Hermitian operator on H_r. The hypothesis (2.1) is fulfilled if we make the assumption

$$\det\{a_{rs}(x_r)\}_{r,s=1}^{k} > 0 \tag{2.5}$$

for all $(x_1,x_2,\ldots,x_k) \in I_k$, the Cartesian product of the k intervals $0 \leq x_r \leq 1$. Theorem 2.1 can now be used to establish that the eigenfunctions of the multi-parameter Sturm-Liouville problem form a complete ortho-normal set with respect to the weight function $\det\{a_{rs}(x_r)\}_{r,s=1}^{k}$ in $L^2[I_k]$. Note that there is no loss of generality in assuming that zero is in the resolvent set of each T_r. This is easily seen if we make use of the following lemma

lemma 2.1 [2] or [15]

A necessary and sufficient condition for $\det\{a_{rs}(x_r)\}_{r,s=1}^{k}$ to be non-zero for all $x \in I_k$ is that for any set $\varepsilon_1,\ldots,\varepsilon_k = \pm 1$ there exists a k-tuple $\mu_1,\ldots,\mu_k$ such that

$$\varepsilon_r \sum_{s=1}^{k} \mu_s\, a_{rs}(x_r) > 0\ , \qquad r = 1,\ldots,k \tag{2.6}$$

for all $x = (x_1,\ldots,x_k) \in I_k$.

If we choose $\varepsilon_r = 1$, $r = 1,\ldots,k$ and shift the λ-origin by writing $\lambda_s = \lambda'_s - \alpha\mu_s$, and defining

$$q_r^+ = q_r + \alpha \sum_{s=1}^{k} \mu_s \, a_{rs}(x_r)$$

where the sum satisfies (2.6) with $\varepsilon_r = 1$ and $\alpha > 0$ so as to make q_r^+ sufficiently large and positive so that the problems

$$\frac{d^2 y_r(x_r)}{dx_r^2} - q_r^+(x_r) y_r(x_r) + \nu_r \, y_r(x_r) = 0$$

together with the boundary conditions (2.4) do not have $\nu_r = 0$ as an eigenvalue.

For the multi-parameter Sturm-Liouville problem we have a fairly well developed oscillation theory [9] and for $k = 2$ there is an oscillation theory for the system (2.2) in the case of non-linear dependence on λ_1 and λ_2 , [8, 11, 12]. Also in the case of two-parameters an alternative treatment of the completeness problem may be found in [14].

Although the "regular" multi-parameter problem as briefly discussed above is now becoming fairly well developed the so called "singular" case remains virtually unexplored and it is to this situation we now turn.

3. A generalised Weyl Limit-point, Limit-circle theory

Let $[a_r, b_r)$, $r = 1,\ldots,k$ denote k semi-open intervals in which a_r is finite and b_r is arbitrary. Assume

$$\begin{aligned} &a_{rs}(x_r) \in C[a_r,b_r) , \qquad r,s = 1,\ldots,k \\ &q_r(x_r) \in C[a_r,b_r) , \qquad r \quad = 1,\ldots,k \end{aligned} \tag{3.1}$$

and that $a_{rs}(x_r)$, $q_r(x_r)$ are real valued. Further it is assumed that (2.5) holds for all $x = (x_1,\ldots,x_k) \in I_k$, where I_k is now used to denote the Cartesian product of the intervals $a_r \leq x_r < b_r$, $r = 1,\ldots,k$. For our Hilbert spaces we take $H_r = L^2_{\rho_r}[a_r,b_r)$, $r = 1,2,\ldots,k$, where the weight function ρ_r is defined below.

We wish to study the solutions of the system

$$-\frac{d^2 y_r}{dx_r^2} + \sum_{s=1}^{k} \{a_{rs}(x_r)\lambda_s + q_r(x_r)\} y_r(x_r) = 0 \ , \quad x_r \in [a_r, b_r) \ . \tag{3.2}$$

The starting point is to consider the boundary condition functions $\theta_r(x_r;\lambda)$, $\varphi_r(x_r;\lambda)$, $r = 1,2,\ldots,k$ where

$$\begin{aligned} \theta_r(a_r;\lambda) &= \cos \alpha_r \ , & \theta'_r(a_r;\lambda) &= \sin \alpha_r \ , \\ \varphi_r(a_r;\lambda) &= \sin \alpha_r \ , & \varphi'_r(a_r;\lambda) &= -\cos \alpha_r \ , \end{aligned} \tag{3.3}$$

where λ is the k-tuple $(\lambda_1,\lambda_2,\ldots,\lambda_k)$, $\alpha_r \in [0,\pi)$ and $' \equiv \frac{d}{dx_r}$. The boundary condition functions are solutions of (3.2) for each r and may be shown to exist by the standard existence theorems for ordinary differential equations.

We now form linear combinations $\psi_r(x_r;\lambda)$ of the boundary condition functions and ask; in what sense do the solutions $\psi_r(x_r;\lambda) \in L^2_{\rho_r}[a_r,b_r)$? Thus we seek to generalise the Weyl limit-point, limit-circle theory from the one-parameter case to the multi-parameter case. Closely related to this question is the problem of determining under what conditions the tensor product

$$\psi_1(x_1,\lambda) \otimes \ldots \otimes \psi_k(x_k;\lambda) \in H \ ,$$

where

$$H = \prod_{r=1}^{k} \otimes H_r \ .$$

At this point we find it convenient to reformulate the system (3.2) in order to introduce some symmetry into the problem and to simplify the subsequent analysis. To do this we would like to claim that lemma 2.1 hold for the conditions (3.1) and the definition of I_k in this section. In particular we would like the lemma to hold when some or all of the ends points a_r are infinite. However, it is easy to find counter examples to show that the lemma is not true in general unless we assume additional hypotheses.

Thus we make the permanent assumption that for any set $\varepsilon_1,\dots,\varepsilon_k = \pm 1$, there exists a k-tuple $(\alpha_1,\alpha_2,\dots,\alpha_k)$ of real numbers, $(\alpha_r \neq 0,\ r = 1,\dots,k)$ such that

$$\rho_r = \sum_{s=1}^{k} \varepsilon_r \alpha_s a_{rs}(x_r) > 0 \tag{3.4}$$

for all $x \in I_k$, $r = 1,\dots,k$.

It will become clear, in the subsequent discussion, that the condition (3.4) is necessary, for if (3.4) does not hold it is possible to find examples of equations of the form (3.2) with (2.5) satisfied for which, for at least one r, no square integrable solutions exist.

By replacing λ_s in (3.2) by $\lambda_s \alpha_s$ and making use of (3.4) the system may be rewritten in the form

$$-\frac{1}{\rho_r} \frac{d^2 y_r}{dx_r^2} + \left\{ \sum_{p=1}^{k} \lambda_p s_{rp}(x_r) + t_r(x_r) \right\} y_r(x_r) = 0 \tag{3.5}$$

where

$$s_{rp} = \frac{\alpha_p a_{rp}(x_r)}{\rho_r}, \qquad t_r = \frac{q_r}{\rho_r},$$

and hence

$$\varepsilon_r \sum_{p=1}^{k} s_{rp}(x_r) = 1 . \tag{3.6}$$

For the sake of simplicity and without loss of generality we restrict the analysis to the case $k = 2$ only. First, consider the system (3.5) $(k = 2)$ in the separate spaces H_1 and H_2, that is consider the ordinary differential equation

$$-\frac{1}{\rho(x)} \frac{d^2 y}{dx^2} + (\lambda_1 s_1(x) + \lambda_2 s_2(x) + t(x))y = 0 , \tag{3.7}$$

$$x \in [a,b) ,$$

where $[a,b)$ is a semi-open interval with a finite and b arbitrary, $S_i(x)$, $(i = 1,2)$ $t(x)$ and $\rho(x)$ (>0) are real valued continuous functions on $[a,b)$ such that

$$S_1(x) + S_2(x) = 1 , \qquad x \in [a,b) . \tag{3.8}$$

(i.e. we assume one of (3.6) to hold with $\varepsilon_r = 1$).

Let $\varphi(x;\lambda_1,\lambda_2)$, $\theta(x;\lambda_1,\lambda_2)$ be boundary condition functions satisfying (3.7) and the initial conditions

$$\varphi(a;\lambda_1,\lambda_2) = \sin \alpha , \qquad \varphi'(a;\lambda_1,\lambda_2) = - \cos \alpha , \tag{3.9}$$

$$\theta(a;\lambda_1,\lambda_2) = \cos \alpha , \qquad \theta'(a;\lambda_1,\lambda_2) = \sin \alpha , \quad \alpha \in [0,\pi) .$$

Finally introduce the set

$$\Lambda = \left\{ \lambda_1,\lambda_2 | I_m(\lambda_1 S_1 + \lambda_2 S_2) \text{ is of one sign and non-zero} \right\} \tag{3.10}$$

$$\text{for all } x \in [a,b) .$$

By virtue of (3.4) it is clear that Λ is non-empty.

The first result towards generalising the limit-point limit-circle theory is contained in the following

<u>Theorem 3.1</u> [15]

Let $[a,b)$ be an interval where a is finite and b is arbitrary. Let $S_1(x)$, $S_2(x)$, $\rho(x) > 0$, $t(x)$ be continuous for all $x \in [a,b)$ and let $\lambda_1,\lambda_2 \in \Lambda$. Then there exists a generalised Weyl coefficient $M(b;\lambda_1,\lambda_2)$ such that

$$\psi(x;\lambda_1,\lambda_2) = \varphi(x;\lambda_1,\lambda_2) \, M(b;\lambda_1,\lambda_2) + \theta(x;\lambda_1,\lambda_2)$$

is a solution of (3.7) having the property that

$$\left| I_m \int_a^b |\psi|^2 (\lambda_1 S_1 + \lambda_2 S_2) \rho(x) dx \right| < \infty . \tag{3.11}$$

If a is a regular point of the equation (3.7) then the following limit-point, limit-circle situations prevail at the end point b. In the limit-circle case there are two linearly independent solutions of (3.7) each of which is square integrable on $[a,b)$ with weight function $\rho|\nu_1 S_1 + \nu_2 S_2|$, $(\nu_i = I_m \lambda_i,\ i = 1,2)$. These solutions are $\varphi(x;\lambda_1,\lambda_2)$ and $\psi(x;\lambda_1,\lambda_2)$ respectively. If a limit-point situation arises at $x = b$ then there is only one linearly independent solution square integrable on $[a,b)$ with weight function $\rho|\nu_1 S_1 + \nu_2 S_2|$. This solution is $\psi(x;\lambda_1,\lambda_2)$. Notice in this case that although $\varphi(x;\lambda_1,\lambda_2)$ is not square integrable in the above sense, it may happen that $\varphi \in L^2_\rho[a,b)$. Nevertheless this is still a limit-point situation. Examples illustrating this possibility are to be found in [15]. From the above discussion we have the following limit-point limit-circle classification.

(I) Limit circle case (LC)

$$\psi,\varphi \in L^2_\omega[a,b), \quad \omega = \rho|\nu_1 S_1 + \nu_2 S_2|, \quad \lambda_1,\lambda_2 \in \Lambda .$$

(II) Limit point case (LP1)

$$\psi \in L^2_\omega[a,b), \quad \varphi \notin L^2_\omega[a,b), \quad \varphi \notin L^2_\rho[a,b),$$

$$\lambda_1\lambda_2 \in \Lambda .$$

(III) Limit point case (LP2)

$$\psi \in L^2_\omega[a,b), \quad \varphi \notin L^2_\omega[a,b), \quad \varphi \in L^2_\rho[a,b),$$

$$\lambda_1\lambda_2 \in \Lambda .$$

Unlike the one-parameter case in which the limit-point, limit-circle cases are independent of the spectral parameter, the multi-parameter case exhibits no such property in general. In fact we have

Theorem 3.2 [15]

Suppose (i) there exists a set Λ^* of real numbers, not identically zero, defined by

$$\Lambda^* = \left\{ \mu_1, \mu_2 \,\middle|\, \mu_1 S_1(x) + \mu_2 S_2(x) > 0 \quad \text{for all} \quad x \in [a,b) \right\}$$

(ii) for every $\mu_1, \mu_2 \in \Lambda^*$ and given complex numbers $\lambda_{01}, \lambda_{02}$ together with corresponding boundary condition functions $\theta(x;\lambda_{01},\lambda_{02})$, $\varphi(x;\lambda_{01},\lambda_{02})$

$$\sum_{r=1}^{2} \int_a^b \mu_r S_r(x) \left|\theta(x;\lambda_{01},\lambda_{02})\right|^2 \rho(x)dx < \infty$$

$$\sum_{r=1}^{2} \int_a^b \mu_r S_r(x) \left|\varphi(x;\lambda_{01},\lambda_{02})\right|^2 \rho(x)dx < \infty .$$

Then for all solutions $\theta(x;\lambda_1,\lambda_2)$, $\varphi(x;\lambda_1,\lambda_2)$ of (3.7), (3.9) with

$$\operatorname{Re}(\lambda_{0r} - \lambda_r), \quad I_m(\lambda_{0r} - \lambda_r) \in \Lambda^*, \qquad r = 1,2 .$$

$$\sum_{r=1}^{2} \int_a^b \mu_r S_r(x) \left|\theta(x;\lambda_1,\lambda_2)\right|^2 \rho(x)dx < \infty ,$$

$$\sum_{r=1}^{2} \int_a^b \mu_r S_r(x) \left|\varphi(x;\lambda_1,\lambda_2)\right|^2 \rho(x)dx < \infty .$$

<u>Corollary 3.1</u>

If (i) $S_r(x) > 0$ $(r = 1,2)$ for all $x \in [a,b)$

(ii) there exist parameters $\lambda_{01}, \lambda_{02}$ and associated solutions $\theta(x;\lambda_{01},\lambda_{02})$, $\varphi(x;\lambda_{01},\lambda_{02})$ such that

$$\int_a^b |\theta(t;\lambda_{01},\lambda_{02})|^2 \rho \, dt < \infty ,$$

$$\int_a^b |\varphi(t;\lambda_{01},\lambda_{02})|^2 \rho \, dt < \infty .$$

Then for all solutions $\theta(x;\lambda_1,\lambda_2)$, $\varphi(x;\lambda_1,\lambda_2)$ we have

$$\int_a^b |\theta(t;\lambda_1,\lambda_2,)|^2 \rho \, dt < \infty ,$$

$$\int_a^b |\varphi(t;\lambda_1,\lambda_2)|^2 \rho \, dt < \infty .$$

From theorem 3.2 we may conclude that if for $\lambda_1 = \lambda_{01}$, $\lambda_2 = \lambda_{02}$ a limit-circle situation prevails at $x = b$, then we always have a limit-circle case for all parameters λ_1, λ_2 provided $\mathrm{Re}(\lambda_{0r} - \lambda_r)$, $I_m(\lambda_{0r} - \lambda_r) \in \Lambda^*$, $r = 1,2$. In the limit-point case we may argue as follows: suppose for $\lambda_1 = \lambda_{01}$, $\lambda_2 = \lambda_{02}$, a limit-point case occurs at $x = b$ and if for another pair $\lambda_{11}, \lambda_{12}$ where the real and imaginary parts of $(\lambda_{0r} - \lambda_{1r})$, $r = 1,2$, are in Λ^* a limit-circle case occurs, then theorem 3.2 would imply that the limit-circle case occurs for $\lambda_1 = \lambda_{01}, \lambda_2 = \lambda_{02}$ also. This is a contradiction and so we conclude that if $\lambda_1 = \lambda_{01}, \lambda_2 = \lambda_{02}$ gives rise to a limit point case, then $\lambda_1 = \lambda_{11}$, $\lambda_2 = \lambda_{12}$ also gives rise to a limit point case. Corollary 3.1 however provides the usual Weyl result, namely that the occurrence of the limit-point case and the limit-circle case respectively is independent of λ_1, λ_2.

4. Limit-point, Limit circle classification in $H_1 \otimes H_2$

Let $[a_i, b_i)$, $(i = 1,2)$ be a semi-open interval with a_i finite and b_i arbitrary. Let $S_i(x_i)$, $t_i(x_i)$, $i = 1,2$ be continuous functions on $[a_i, b_i)$ such that

$$S_i(x_i) + t_i(x_i) = 1 , \quad \text{for all } x_i \in [a_i, b_i) , \quad i = 1,2, ,$$

and (2.5) holds.

Let $\rho_i(x_i) > 0$, $r_i(x_i)$ be continuous for all $x_i \in [a_i, b_i)$. Denote by $\varphi_i(x_i;\lambda_1,\lambda_2)$, $\theta_i(x_i;\lambda_1,\lambda_2)$ respectively, solutions of

$$-\frac{1}{\rho_i(x_i)}\frac{d^2 y_i}{dx_i^2} + \left\{ \lambda_1 S_i(x_i) + \lambda_2 t_i(x_i) + r_i(x_i) \right\} y_i = 0 , \tag{4.1}$$

satisfying the initial conditions

$$\begin{aligned} \varphi_i(a_i;\lambda_1,\lambda_2) &= \sin \alpha_i , & \varphi_i'(a_i;\lambda_1,\lambda_2) &= -\cos \alpha_i , \\ \theta_i(a_i;\lambda_1,\lambda_2) &= \cos \alpha_i , & \theta_i'(a_i;\lambda_1,\lambda_2) &= \sin \alpha_i , \end{aligned} \tag{4.2}$$

in which $\alpha_i (i = 1,2)$ is real and $\alpha_i \in [0,\pi)$.

Corresponding to theorem 3.1 we have the result

Theorem 4.1 [15]

There exists a solution pair

$$\psi_i(x_i;\lambda_1,\lambda_2) = \varphi_i(x_i;\lambda_1,\lambda_2)\, M_i(\lambda_1,\lambda_2) + \theta_i(x_i;\lambda_i,\lambda_2) ,$$

$(i = 1,2)$ of (4.1) having the property that

$$\int_{a_1}^{b_1}\int_{a_2}^{b_2} |\psi_1\, \psi_2|^2\, (S_1\, t_2 - S_2\, t_1)\rho_1\, \rho_2\, dx_1\, dx_2 < \infty . \qquad (4.3)$$

We now turn to the problem of limit-point limit-circle classification in $H_1 \otimes H_2$. Firstly it is easy to prove that if a limit-circle situation prevails in both H_1 and H_2 separately then a limit-circle situation prevails in $H_1 \otimes H_2$, in the sense that there exists a product solution $\psi_1\, \psi_2$ satisfying (4.3) as well as a product solution $\varphi_1\, \varphi_2$ also satisfying (4.3). However if a limit point case arises in either or both H_1 and H_2 it is not clear whether a limit point or a limit circle situation prevails in $H_1 \otimes H_2$. Indeed this problem remains largely uninvestigated. For a more detailed account we refer to [15]. Another problem of considerable interest is to ask whether theorem 3.2 has an analogue in the tensor product space. If such a result can be proved we conjecture that the limit-point, limit-circle classification in $H_1 \otimes H_2$ is independent of the parameters λ_1, λ_2.

Before we leave this aspect of the multi-parameter eigenvalue problem we mention that an "operator" formulation of the problem may be given in which we construct generalised Weyl subspaces and consider (3.7) or its multi-parameter analogue on these spaces. For such operators defined on these domains we may prove that the differential operators are J- self adjoint in the sense of Glazman [7]. Also, after a modification of the work by Evans [6] we may obtain sufficient conditions for the occurrence of the limit-point limit-circle cases described in section 3 above. These results will be reported elsewhere.

As far as completeness and expansion theorems are concerned we draw attention to the recent paper of Browne [4] in which a "limiting" parseval equality is obtained. Browne's approach is different from ours in the sense that he considered the Sturm-Liouville system defined by (3.2) and the boundary conditions

$$\begin{aligned} y_r(a_r)\cos\alpha_r + y_r'(a_r)\sin\alpha_r &= 0\,, \qquad 0 < \alpha_r \leq \pi\,, \\ y_r(b_r)\cos\beta_r + y_r'(b_r)\sin\beta_r &= 0\,, \qquad 0 \leq \beta_r < \pi\,, \end{aligned} \tag{4.4}$$

and examined the system when some or all of the variables b_r are allowed to range over a half line $[a_r,\infty)$.

5. Direct formulation of the Problem in the Tensor Product Space

Despite the success of treating multi-parameter spectral problems, first in the separate Hilbert spaces H_r and then in the tensor product space, it seems natural that perhaps the most powerful results are to be obtained by formulating the problem directly in the tensor product space. There are several ways of doing this; either by using an integral equation formulation [13] or by using a method of separation of "variables" in reverse [15] to eliminate all but one of the spectral parameters λ_r $(r = 1,\ldots,k)$ and hence to consider a system of spectral problems of the form

$$S_r\, u = \lambda_r\, Tu\,, \qquad (r = 1,2,\ldots,k) \tag{5.1}$$

in which S_r is a partial differential operator, not necessarily elliptic, and T denotes multiplication by a positive function. This approach has received little attention. However see the work of Cordes [5] in which a restricted form of the two-parameter case is treated in detail.

References

[1] Atkinson, F. V.: Multi-parameter spectral theory, Bull. Amer. Math Soc. 74 1-28 (1968).

[2] Atkinson, F. V.: Multi-parameter eigenvalue problems Vol. 1, Academic Press (1972).

[3] Browne, P. J.: A multi-parameter eigenvalue problem. J. Math. Anal. and Applications 38 553-568 (1972).

[4] Browne, P. J.: A singular multi-parameter eigenvalue problem in second order ordinary differential equations. J. Diff. equations 12 81-94 (1972).

[5] Cordes, H. O.: Über die spektralzerlegung von hypermaximalen operatoren die durch separaten der variablen zerfallen. I. Math Ann 128 257-289 (1954). II Math Ann 129 373-411 (1955).

[6] Evans, W. D.: On the limit-point, limit-circle classification of a second-order differential equation with a complex coefficient. J. Lond. Math. Soc (2) 4 245-256 (1971).

[7] Glazman, I. M.: Direct methods of qualitiative spectral analysis of singular differential operators, Israel Program for Scientific Translations (Jerusalem, 1965).

[8] Greques, M., Neuman, F., Arscott, F. M.: Three-point-boundary value problems in differential equations. J. Lond. Math. Soc. 3 429-436 (1971).

[9] Ince, E. L.: Ordinary differential equations. Longmans Green (1926).

[10] Sleeman, B. D.: Multi-parameter eigenvalue problems in ordinary differential equations. Bul. Inst. Poli. Jassy 17 (21) 51-60 (1971).

[11] Sleeman, B. D.: The two-parameter Sturm-Liouville problem for ordinary differential equations. Proc. Roy. Soc., Edinburgh Sect. A 69 139-148 (1971).

[12] Sleeman, B. D.: The two-parameter Sturm-Liouville problem for ordinary differential equations, II, Proc. Amer. Math. Soc. 34 165-170 (1972).

[13] Sleeman, B. D.: Multi-Parameter Eigenvalue Problems and k-linear Operators. Conference on the theory of ordinary and partial differential equations. Lecture Notes in Mathematics Vol. (280) 347-353. Springer (1972).

[14] Sleeman, B. D.: Completeness and Expansion Theorems for a Two-parameter Eigenvalue Problem in Ordinary Differential Equations using Variational Principles. J. Lond. Math. Soc. (2) 6 705-712 (1973).

[15] Sleeman, B. D.: Singular-Linear Differential Operators with Many Parameters. Proc. Roy. Soc. Edinburgh Sec. A. 71 199-232 (1973).

INTEGRAL - EQUATION FORMULATION OF TWO-PARAMETER EIGENVALUE PROBLEMS

F. M. ARSCOTT
Department of Mathematics
University of Reading, England

1. Introduction

The problem considered in this paper is a two-parameter eigenvalue problem in differential equations of the form

$$\frac{d^2w}{dz^2} + (\lambda + \mu f(z) + g(z))\, w = 0, \tag{1.1a}$$

associated with which are the three-point boundary conditions

$$w(a) = w(b) = w(c) = 0. \tag{1.1b}$$

Here $f(z)$,$g(z)$ are prescribed functions, while the parameters λ,μ are at our disposal and have to be chosen so that (1.1a) has a non-trivial solution satisfying (1.1b). It is convenient to take the variables in (1.1a) as complex, since this is the case in many problems of practical importance (Arscott (1964a, 1964b)). In some problems, one or more of the conditions in (1.1b) is replaced by the condition of vanishing derivative; the theory can be extended to cover such problems with only minor modifications, so we shall confine ourselves to conditions (1.1b).

The main purpose of this paper is to show how the problem (1.1) can be transformed into an integral-equation problem. This is, of course, a very familiar procedure, generally carried out so as to result in an integral equation whose nucleus is a Green's function G or Neumann's function N. Here, however, we shall be concerned with integral equations in which the nucleus is an analogue of the kernel function. Such a nucleus is more difficult to construct, in principle, than the Green's or Neumann function, since it arises as a solution of a partial differential equation; nevertheless it has two compensating advantages. In the first place, our nucleus has smoothness properties which the Green's function lacks, and in the second place it is less stringently dependent on the boundary conditions. As a result, we obtain not merely integral equations which are satisfied by solutions of (1.1), but also integral relations between different solutions of (1.1a): such relations prove extremely valuable in applications. Though made in the context of partial differential equations, a remark by Garabedian is appropriate here: "Because it has no physical significance the kernel function is of more recent origin than the Green's and Neumann's functions. However, in certain respects it provides a more effective tool for the solution of boundary value problems". (Garabedian (1964) p.256).

This paper is concerned with formal theory only; existence problems have been considered in Greguš (1971) and in several papers by Sleeman.

2. A one-parameter problem

It is useful first to formulate a theorem relating to the familiar one-parameter Sturm-Liouville problem

$$\frac{d^2w}{dz^2} + (\lambda - q(z))w = 0 \tag{2.1a}$$

i.e. $$L_z(w) \equiv \left(\frac{d^2}{dz^2} - q(z)\right) w = -\lambda w$$

with boundary conditions

$$w(a) = w(b) = 0. \tag{2.1b}$$

The function $q(z)$ is assumed to be sufficiently well-behaved, e.g. continuous on $[a,b]$

Let the eigenvalues of (2.1) be denoted by λ_i and corresponding eigenfunctions by $w_i(z)$. Then we have the first integral relation:

<u>Theorem 1.</u>

Let (i) $w(z)$ be a solution of (2.1a),

(ii) $K(z,\zeta)$ be a solution of $L_zK - L_\zeta K = 0$, i.e.

$$\frac{\partial^2 K}{\partial z^2} - \frac{\partial^2 K}{\partial \zeta^2} = [q(z) - q(\zeta)]K, \tag{2.2}$$

K being analytic when z,ζ belong to suitable domains D_z, D_ζ in the complex z,ζ planes,

(iii) C_ζ be a path in D_ζ such that

$$[K(z,\zeta)w'(\zeta) - w(\zeta)\, K_\zeta(z,\zeta)]_{C_\zeta} = 0 \tag{2.3}$$

(the symbol $[\;]_{C_\zeta}$ denoting, naturally, the difference in the values at the two ends of C_ζ),

(iv) the integral

$$W(z) \equiv \int_{C_\zeta} K(z,\zeta)\, w(\zeta)\, d\zeta \tag{2.4}$$

exist for all $z \in D_z$ and, if singular, be twice differentiable with respect to z in D_z.

Then $W(z)$ satisfies (2.1) for all $z \in D_z$.

The proof involves only straightforward manipulation and is given in Arscott (1964b).

We observe that the theorem, as it stands, is concerned only with solutions of the differential equation, and no reference is made to the associated boundary conditions. Thus $w(z)$ is required only to be a solution of (2.1a); note also that we assert only that $W(z)$ satisfies this equation; it is not necessarily the same as $w(z)$, or even a multiple thereof; it may indeed be merely the identically zero solution.

However, a chain of corollaries brings out the application of this theorem to the boundary-value problem. Their verification is simple.

Corollary 1. Let $w(z)$ also satisfy (2.1b), i.e. $w = w_i(z)$ for some i. Let $K(z,\zeta)$ as a function of ζ, satisfy (2.1b) also. Then condition (2.4) is satisfied automatically for any C_ζ joining $\zeta = a$, $\zeta = b$ and avoiding any singularity of the differential equation.

Corollary 2. Let, further, $K(z,\zeta)$ also satisfy (2.1b) as a function of z (which will certainly be the case if $K(z,\zeta) = K(\zeta,z)$). Then $W(z)$ is either identically zero or an eigenfunction corresponding to λ_i.

Corollary 3. Let, further, λ_i be a simple eigenvalue, so that $w_i(z)$, $w_j(z)$ are linearly independent for $j \neq i$.
Then $W(z) = \Lambda\, w_i(z)$ for some Λ, possibly 0, i.e. $w_i(z)$ satisfies the integral equation

$$\int_{C_\zeta} K(z,\zeta)\, w_i(\zeta)\, d\,\zeta = \Lambda w_i(z) \qquad (2.5)$$

The genesis of such integral equations seems to lie in the work of Whittaker (1914). Among later developers must be mentioned particularly von Koppenfels (1936) and, more recently Sips((1961and other papers). Much of this development has been made in the context of particular equations, notably those of Mathieu and Lamé. Correspondence between eigenfunctions of (2.1) and of (2.5) may be quite involved; von Koppenfels pointed out that multiple eigenvalues of (2.1) may be separated out into simple eigenvalues of (2.5); on the other hand, some eigenfunction of (2.1) may become 'lost' in (2.5) in the sense that they correspond to $\Lambda = 0$.

Among applications of (2.4) and 2.5) may be mentioned the following

(i) the behaviour of $w(z)$ for certain domains of z, in terms of its behaviour in other domains (e.g., C_ζ may lie near the origin, but $|z|$ be large).

(ii) if $q(z)$ involves a parameter, say α, then $K(z,\zeta)$ involves α also and useful approximation may be possible for certain values of α.

(iii)by choosing $K(z,\zeta)$ as <u>unsymmetric</u>, we may be able to construct a second solution of (2.1a) in terms of the first solution, $w(z)$.

(iv) a bilinear expansion of the nucleus $K(z,\zeta)$ is often possible and is important in physical applications; thus if K in equation (2.5) is sufficiently well-behaved, as a function of ζ, to have an expansion in series of the (suitably normalised) eigenfunctions $w_i(\zeta)$, then it has the bilinear expansion

$$K(z,\zeta) = \sum_i \Lambda_i\, w_i(z)\, w_i(\zeta). \qquad (2.6)$$

Finally, mention should be made of the possibility of iterating the integral operation of Theorem 1; this technique has been much used by Sips (1961)*

* I am indebted to Dr. Kathleen M. Urwin for this reference.

3. A discrete analogue

If the problem (2.1) is replaced by an analogous difference equation, then the following emerges as a parallel to Theorem 1:-

<u>Theorem 2</u>.

Let A be a symmetric matrix with eigenvalues λ_i, eigenvectors x_i.
Let K be any matrix that commutes with A.
Then $\xi_i \equiv Kx_i$ is also an eigenvector of A corresponding to λ_i.

Proof: $$A\xi_i = AKx_i = KAx_i = K\lambda x_i = \lambda Kx_i = \lambda\xi_i.$$

Corollary: If λ_i is a simple eigenvalue of A then ξ_i is a multiple of x_i, say $\xi_i = \Lambda_i x_i$, hence $Kx_i = \Lambda x_i$. Thus x_i is also an eigenvector of K

but associated with a different eigenvalue.

4. The two-parameter eigenvalue problem - integral equations of Malurkar type

Turning now to the problem posed by (1.1a,b), it is convenient to denote the eigenvalue pairs by ordering first μ, then λ, thus writing

$$\mu = \mu_i, \ (i = 0,1,2,\ldots),$$
$$\left.\begin{aligned}\lambda &= \lambda_{ij}\\ w &= w_{ij}\end{aligned}\right\}(i = 0,1,2,\ldots,\ j = 0,1,2,\ldots), \qquad (4.1)$$

(for given i, the enumeration over j is not necessarily infinite; i.e. μ_i having been fixed, there may be only a finite number, perhaps only one, of corresponding λ).

Since (1.1) is a two-parameter problem, an obvious line of attack is reducing it to a one-parameter problem. Naturally, by putting $q(z) = -\mu f(z) - g(z)$, we may apply Theorem 1 and so obtain an integral equation for $w_{ij}(z)$ of the form

$$\Lambda_{ij}\, w_{ij}(z) = \int_a^b K(z,\zeta)\, w_{ij}(\zeta)\, d\zeta \qquad (4.2)$$

(or a similar relation with integration over (b,c)), but this does not constitute a true reduction to a one-parameter problem since λ has been replaced by Λ and the parameter μ occurs in the partial differential equation which K must satisfy. A genuine reduction is made in the following theorem, however:

Theorem 3.

Let (i) $w(z)$ satisfy (1.1a)

(ii) $H(\alpha,\beta,\gamma)$ satisfy the partial differential equation

$$\sum_{\alpha,\beta,\gamma} \left\{f(\beta) - f(\gamma)\right\} \frac{\partial^2 H}{\partial\alpha^2} = -H \sum_{\alpha,\beta,\gamma} g(\alpha)\left\{f(\beta) - f(\gamma)\right\}, \qquad (4.3)$$

$H(\alpha,\beta,\gamma)$ being analytic functions of α,β,γ in domains $D_\alpha, D_\beta, D_\gamma$,

(iii) C_α, C_β be paths in D_α, D_β such that

$$\left[H_\alpha w(\alpha) - H\, w'(\alpha)\right]_{C_\alpha} = \left[H_\beta w(\beta) - H\, w'(\beta)\right]_{C_\beta} = 0 \qquad (4.4)$$

(iv) $$W(\gamma) \equiv \int_{C_\alpha}\int_{C_\beta} H(\alpha,\beta,\gamma)\, w(\alpha) w(\beta)\left\{f(\alpha) - f(\beta)\right\} d\alpha\, d\beta \qquad (4.5)$$

exist for all γ in D_γ and, if singular, converge uniformly with respect to γ in D_γ.

Then $W(z)$ is a solution of (1.1a).

The proof again involves only manipulation and is given in Arscott (1964b).

Similar corollaries may be enunciated as for Theorem 1;

(1). If $C_\alpha = (a,b)$, $C_\beta = (b,c)$, w satisfies all the conditions (1.1b), i.e. $w = w_{ij}(z)$, $H(\alpha,\beta,\gamma)$ as a function of α satisfies the boundary conditions at a,b and, as a function of β, satisfies the conditions at b,c, then relations (4.4) are satisfied automatically.

(2). If H also satisfies all the boundary conditions, as a function of γ, then $W(\gamma) = \Lambda_{ij}\, w_{ij}(\gamma)$, with Λ_{ij} possibly zero

(3). If (λ_{ij},μ_i) is a simple eigenvalue pair, then w_{ij} satisfies the integral equation

$$\Lambda_{ij}\, w_{ij}(\gamma) = \int_{C_\alpha}\int_{C_\beta} H(\alpha,\beta,\gamma)\, w_{ij}(\alpha) w_{ij}(\beta) \left\{f(\alpha)-f(\beta)\right\} d\alpha d\beta \tag{4.6}$$

Relations (4.5),(4.6) are capable of similar applications to those of (2.4),(2.5) described in § 2 above. The expansion of $H(\alpha,\beta,\gamma)$ is now of the form

$$H(\alpha,\beta,\gamma) = \sum_i \sum_j \Lambda_{ij}\, w_{ij}(\alpha)\, w_{ij}(\beta)\, w_{ij}(\gamma) \tag{4.7}$$

provided the w_{ij} are suitably normalised.

It is noteworthy that we have now succeeded in obtaining a genuine one-parameter problem, since the partial differential equation to be satisfied by H is independent of λ and μ, and there remains only the parameter Λ. However, the integral equation (4.7) is now non-linear - an unexpected feature in an essentially linear problem. Such equations seem to appear first in the work of Malurkar (1935).

A word is appropriate here regarding the usefulness of Theorems 1 and 3. In each instance, it appears that transformation of the original differential-equation problem into integral-equation form involves the solution of a partial differential equation, which is an essentially harder problem, so that nothing has been gained. This is not wholly true, even in the general case, because we impose only mild restrictions on the solution of the partial differential equation. More significantly, in the various physical applications which have given rise to two-parameter problems the partial differential equation for K or H is a transformation of a well-known equation such as that of Laplace or Helmholtz, for which many convenient solutions are known. The ordinary differential equations (1.1a),(2.1a) arise, indeed, from precisely the associated partial differential equations (2.2),(4.3) by separation of variables. (Arscott (1964a), ch. I)

5. Whittaker-McLachlan type integral relations

A different type of integral relation, closely connected with the two-parameter problem (1.1), appears in McLachlan (1947), § 10.50 for the particular case of Mathieu's equation; close examination shows that the result quoted derives from the general expression for solutions of the wave equation and due to Whittaker (1940). The following generalisation to problems of the type (1.1) appears to be new. The resulting integral relationship involves only a single integral, but the nucleus has to satisfy <u>two</u> partial differential equations.

<u>Theorem 4.</u>

Let (i) $w(z)$ satisfy (1.1a),

(ii) $J(\alpha,\beta,\gamma)$ be analytic in α,β,γ in appropriate domains $D_\alpha, D_\beta, D_\gamma$ and satisfy the two partial differential equations

$$J_{\alpha\alpha} + \left\{\mu f(\alpha) + g(\alpha)\right\} J = J_{\beta\beta} + \left\{\mu f(\beta) + g(\beta)\right\} J = J_{\gamma\gamma} + \left\{\mu f(\gamma) + g(\gamma)\right\} J, \tag{5.1}$$

(iii) C_γ lie in D_γ and be such that $\left[J_\gamma\, w(\gamma) - J\, w'(\gamma)\right]_{C_\gamma} = 0,$ (5.2)

(iv) $F(\alpha,\beta)$ be defined by

$$F(\alpha,\beta) = \int_{C_\gamma} J(\alpha,\beta,\gamma)\, w(\gamma)\, d\gamma, \tag{5.3}$$

Then $F(\alpha,\beta)$ satisfies the two equations

$$\frac{\partial^2 F}{\partial\alpha^2} + \left\{\lambda + \mu f(\alpha) + g(\alpha)\right\} F = 0,$$

$$\frac{\partial^2 F}{\partial\beta^2} + \left\{\lambda + \mu f(\beta) + g(\beta)\right\} F = 0. \tag{5.4}$$

Proof: We have

$$F_{\alpha\alpha} + \left\{\mu f(\alpha) + g(\alpha)\right\} F = \int_{C_\alpha} \left[J_{\alpha\alpha} + (\mu f(\alpha) + g(\alpha))J\right] w(\gamma)\, d\gamma$$

$$= \int_{C_\gamma} (J_{\gamma\gamma} + (\mu f(\gamma) + g(\gamma))\, J\, w(\gamma)\, d\gamma \quad \text{(by (5.1)}$$

$$= \left[J_\gamma w(\gamma) - Jw'(\gamma)\right]_{C_\gamma} + \int_{C_\gamma} J\left\{w''(\gamma) + (\mu f(\gamma) + g(\gamma))\ w(\gamma)\right\} d\gamma$$

(on integrating by parts twice)

$$= -\lambda \int_{C_\gamma} J\, w\, d\gamma = -\lambda F.$$

This is the first of equations (5.4) and the second is obtained similarly.

It is interesting to observe that for any sufficiently well-behaved w, the function $F(\alpha,\beta)$ given by (5.3) is a solution of the partial differential equation

$$\frac{\partial^2 F}{\partial\alpha^2} - \frac{\partial^2 F}{\partial\beta^2} + \left[\mu\left\{f(\alpha) - f(\beta)\right\} + g(\alpha) - g(\beta)\right] F = 0; \tag{5.5}$$

it is the condition that w satisfies (1.1a) which makes $F(\alpha,\beta)$ satisfy the two equations (5.4). In special circumstances, $F(\alpha,\beta)$ is actually separable and an integral equation results, as follows:-

Corollary. Let $w = w_{ij}(z)$, an eigenfunction belonging to a <u>simple</u> eigenvalue pair $\lambda = \lambda_{ij}$, $\mu = \mu_{ij}$. Let J, as a function of α, satisfy the boundary conditions (1.1b) at a,b and, as a function of β, satisfy these conditions at b,c, i.e.

$$J(a,\beta,\gamma) = J(b,\beta,\gamma) = J(\alpha,b,\gamma) = J(\alpha,c,\gamma) = 0. \tag{5.6}$$

Then $F(\alpha,\beta) = \Lambda_{ij} w_{ij}(\alpha)\, w_{ij}(\beta)$ for some constant Λ_{ij}, possibly zero.

Thus

$$\Lambda_{ij}\, w_{ij}(\alpha)\, w_{ij}(\beta) = \int_{C_\gamma} J(\alpha,\beta,\gamma)\, w_{ij}(\gamma)\, d\gamma. \tag{5.7}$$

This holds since the conditions imposed make $F(\alpha,\beta)$, as a function of α, a solution of both (1.1a) and (1.1b) and thus either identically zero or a multiple of $w_{ij}(\alpha)$; similarly it is a multiple of $w_{ij}(\beta)$.

There is an interesting symmetry between the two integral equations (4.6) and (5.7); in the former, a single unknown function w is expressed as an integral involving a product of two such functions, while in the latter a product of two functions is expressed as an integral involving a single function. This leads naturally to the conjecture that the two equations are in some way related but if this is so, the connection does not appear to be obvious. It is worth remarking that the nucleus $J(\alpha,\beta,\gamma)$ of (5.7), satisfying the two equations (5.1), also satisfies the equation (4.3) for nuclei of (4.6), but the converse is not true.

6. Möglich-type integral relations

Mention should finally be made of yet another integral relation connected with the two-parameter problem (1.1); this type of relation was formulated by Möglich (1927) in connection with the ellipsoidal wave equation, and used extensively by him in his investigation of the latter. In essence, this is a relation between products of two eigenfunctions; the nucleus is more involved than those for Malurkar-type integral equations, in the sense that it must satisfy a more complicated partial differential equation; on the other hand, the resulting integral equation is linear. The statement of this result, in its general form, appears to be new.

Theorem 5.

Let (i) $W(z,\zeta)$ satisfy the equation

$$M_{z,\zeta}\, W \equiv \frac{\partial^2 W}{\partial z^2} - \frac{\partial^2 W}{\partial \zeta^2} + \left[\mu\left\{f(z) - f(\zeta)\right\} + g(z) - g(\zeta)\right] W = 0, \qquad (6.1)$$

(ii) $U(z,\zeta,z',\zeta')$ be analytic and satisfy $M_{z,\zeta}\, U = M_{z',\zeta'}\, U$ (6.2)

in suitable domains $D_{z,\zeta}$, $D_{z',\zeta'}$.

(iii) $C_{z'}$, $C_{\zeta'}$ be paths in $D_{z',\zeta'}$ such that

$$\left[\frac{\partial U}{\partial z'} W(z',\zeta') - U\frac{\partial W}{\partial z'}\right]_{C_{z'}} = \left[\frac{\partial U}{\partial \zeta'} W(z',\zeta') - U\frac{\partial W}{\partial \zeta'}\right]_{C_{\zeta'}} = 0. \qquad (6.3)$$

(iv)

$$W^*(z,\zeta) = \int_{C_{z'}}\int_{C_{\zeta'}} U(z,\zeta,z',\zeta')\, W(z',\zeta')\left\{f(z') - f(\zeta')\right\} dz'\, d\zeta' \qquad (6.4)$$

exist and be twice differentiable with respect to z,ζ in $D_{z,\zeta}$.

Then $W^*(z,\zeta)$ also satisfies (6.1) in $D_{z',\zeta'}$.

The proof of this again involves only manipulation, though of considerable complexity, and is therefore omitted.

This theorem, since it relates entirely to solutions of a partial differential equation, is of a much more general character than those considered earlier in this paper, and some specialisation is needed before results appear which have immediate application; these may be indicated very briefly:-

(1). Let $W(z',\zeta')$ be taken as the product of two eigenfunctions $w_{ij}(z')w_{ij}(\zeta')$; then if U, regarded as a function of z, ζ is characterised appropriately one can show that $W^*(z,\zeta)$ is also a product of two eigenfunctions, so that (6.4) becomes an integral equation for products.

(2). By giving ζ an appropriate constant value, one derives an integral equation of similar form to the Malurkar-type equation (para 4) in that we have a single function outside the integral and a product of two functions within. Closer examination shows, however, that Malurkar-type equations are not, in general, obtainable from Möglich-type equations in this way; a study of the relationships, for the particular case of the ellipsoidal wave equation, is given in Arscott (1957).

7. Discrete analogue of the two-parameter problem

In paragraph 3 above we discussed briefly the discrete analogue of the one-parameter problem; a discrete analogue of the two-parameter problem can be shown to take the form of simultaneous equations

$$\begin{aligned} Ax &= (\lambda I + \mu P_1 + P_2)\, x, \\ Ay &= (\lambda I + \mu Q_1 + Q_2)\, y, \end{aligned} \tag{7.1}$$

where A is a symmetric matrix, P_1, P_2, Q_1, Q_2 are diagonal matrices, and λ, μ are to be so chosen that equations (7.1) have non-trivial solutions x,y. It seems reasonable to expect that these equations can be attacked and transformed by a process analogous to the conversion of differential-equation problems into integral-equation problems. This, however, seems to require evolutions of a more complicated nature than standard matrix analysis and is a problem calling for careful study and possibly a novel approach using tensors rather than matrices.

REFERENCES

Arscott (1957) Integral equations for ellipsoidal functions, Quart.J. Math. 8, 223.

Arscott (1964a) Periodic Differential Equations, Pergamon Press.

Arscott (1964b) Two-parameter eigenvalue problems in differential equations, Proc.Lond.Math.Soc. 14, 459.

Garabedian (1964) Partial Differential Equations, John Wiley.

Greguš, Neuman, Arscott (1971) Three-point boundary value problems in differential equations, Jour.Lond-Math.Soc. 3, 429.

Malurkar (1935) Ellipsoidal wave functions, Ind.Jour.Phys. 9, 45.

McLachlan (1947) Theory and Application of Mathieu Functions, Oxford.

Möglich (1927) Beugunserscheinungen an Körpern von Ellipsoidischer Gestalt, Ann.d.Phys. 83, 609.

Sips (1953) Recherches sur les fonctions de Mathieu, Bull.Soc.Roy.Sci. de Liège, 444.

Sleeman - Paper in these proceedings.

von Koppenfels (1936) Beitrag zur Theorie der linearen Differentialgleichungen, Math.Ann 112, 24.

Whittaker and Watson (1940) Modern Analysis, Cambridge.

ON DIFFERENTIAL OPERATORS SINGULAR AT THE BOUNDARY

by

R. MARTINI

1. INTRODUCTION

In this lecture we shall deal mainly with differential expressions of the type

$$(1.1)\qquad \alpha D^2 + \beta D$$

on a bounded open interval I of the real axis. Under certain conditions concerning the coefficients α,β (e.g., α is assumed to be positive on I and to vanish at the boundary of I) we shall prove that with the expression (1.1) there can be associated an infinitesimal generator of a strongly continuous contraction semi-group of operators of class (C_o) on the Banach space $C([0,1])$. See for the main result theorem 2.

2. SOME DEFINITIONS

Let X be a Banach space. Then a family $\{T_t\}(t \geq 0)$ of continuous operators on X is a <u>semi-group</u> if

i) $T_{t+s} = T_t . T_s$

ii) $T_o = I$

for all $t > 0$, $s > 0$ and where I is the identity operator on X.

The semi-group is said to be <u>strongly continuous</u> if the map $t \to T_t$ is continuous when the space of continuous linear operators L(X) on X is equipped with the strong operator topology; i.e., $\lim_{s \to t} T_s f = T_t f$ for each $f \in X$ and each $t \in [0, \infty)$.

With a strongly continuous semi-group there is associated an operator called the <u>infinitesimal generator</u>, defined in the following way

$$D(A) = \{f \in X \mid \lim_{t \to 0} \tfrac{1}{t}(T_t f - f) \text{ exists}\}$$

and then

$$Af = \lim_{t \to 0} \frac{1}{t} (T_t f - f)$$

for each $f \in D(A)$.

3. MAIN RESULT

An infinitesimal generator of a strongly continuous semi-group is completely characterized by the classical theorem of Hille-Yosida, which we formulate in the version we need; namely, for contraction semi-groups.

Theorem 1. Let X be a Banach space. Then an operator A with $D(A) \subset X$, $R(A) \subset X$ is an infinitesimal generator of a uniquely determined semi-group $\{T_t\}$ $(t \geq 0)$ if and only if

i) $D(A)$ is dense in X.
ii) The resolvent $R(\lambda;A) = (\lambda - A)^{-1}$ exists with domain X and satisfies $||\lambda R(\lambda;A)|| \leq 1$ for all λ sufficiently large.

Proof. See e.g. YOSIDA [3], p. 248-249.

Concerning infinitesimal generators and differential expressions we have the following theorem.

Theorem 2. Let I be a bounded open interval of the real axis, $\bar{I}$ its closure, X the real space $C(\bar{I})$. Then the operator A induced by the differential expression

$$\alpha D^2 + \beta D \;,$$

where

a) α, β are real functions belonging to $C(\bar{I}) \cap C^2(I)$.
b) $\alpha(x) > 0$ when $x \in I$, $\alpha(x) = 0$ when $x \in \partial I$.
c) α^{-1} is not integrable over neighbourhoods of endpoints of $\bar{I}$.
d) $\beta . \alpha^{-\frac{1}{2}}$ is bounded on I.

on the linear manifold

$$D(A) = \{ f \in C(\bar{I}) | f|\, I \in C^2(I), \lim_{x \to \partial I} A(f|I) = 0 \}$$

is an infinitesimal generator of a strongly continuous contraction semi-group.

Proof. This theorem is proved by means of the theorem of Hille-Yosida. We divide the proof in several steps.

Step 1. Since $C^2(\bar{I})$ is contained in $D(A)$ it follows that $D(A)$ is dense in $C(\bar{I})$. Hence condition i) of the theorem of Hille-Yosida is fullfilled.

Step 2. We have for each $f \in D(A)$, each $x \in \bar{I}$ and each $\lambda > 0$ the inequalities

$$(3.1) \qquad \min_{x \in \bar{I}} \{\lambda f(x) - (Af)(x)\} \leqq \lambda f(x) \leqq \max_{x \in \bar{I}} \{\lambda f(x) - (Af)(x)\}.$$

Since Af vanishes at the boundary ∂I of I it follows that the second part of the inequality (3.1) is valid in case f attains its maximum at a point of the boundary. In case f attains its maximum at an interior point x_0 we have

$$\max_{x \in \bar{I}} \{\lambda f(x) - (Af)(x)\} \geqq \lambda f(x_0) - (Af)(x_0) =$$
$$= \lambda f(x_0) - \alpha(x_0)D^2 f(x_0) - \beta(x_0)Df(x_0) \geqq \lambda f(x_0)$$

for in this case at this point we have $Df(x_0) = 0$ and $D^2 f(x_0) \leqq 0$. We can prove the first part of inequality (3.1) in similar way.

Step 3. From step 2 it follows that for $\lambda > 0$ $\lambda - A$ is injective on $D(A)$ and in this case we have $||\lambda f|| \leqq ||(\lambda - A)f||$. Now A is a closed operator. Hence if the range $R(\lambda;A)$ is dense in $C(\bar{I})$ it follows that $||\lambda R(\lambda;A)|| \leqq 1$. This completes the proof of step 3.

Thus what remains to be proved is the fact that for all λ sufficiently large the range $R(\lambda - A)$ is dense in $C(\bar{I})$. This is in fact the difficult part of the proof of theorem 2. To prove this we must escape from the hard Banach space $C(\bar{I})$ into soft Hilbert spaces. So let us discuss first some Hilbert spaces.

4. SOME SPECIAL HILBERT SPACES.

$H(I,\alpha)$ is the space of all complex-valued functions which are square-integrable over I with respect to the measure $\frac{dx}{\alpha}$. This space is normed by

$$||u||_H = (\int_I |u|^2 \frac{dx}{\alpha})^{\frac{1}{2}}$$

Remark. $H(I,\alpha)$ is a Hilbert space. The space $C_0^\infty(I)$ of smooth functions with compact support in I is dense in $H(I,\alpha)$.

$V(I,\alpha)$ is the space of all $f \in H(I,\alpha)$ with the property that its distributional derivative Df is square-integrable over I with respect to the ordinary Lebesgue measure. This space is normed by

$$||u||_V = (\int_I |u|^2 \frac{dx}{\alpha} + \int_I |Du|^2 dx)^{\frac{1}{2}} .$$

Remark. $V(I,\alpha)$ is a Hilbert space, $C_0^\infty(I)$ is dense in $V(I,\alpha)$ and members of $V(I,\alpha)$ are continuous on $\bar{I}$ and vanishes at the boundary of I. Now it can easily be seen that $V(I,\alpha)$ is a dense subset of $H(I,\alpha)$ with continuous injection J.

Moreover we can prove that J is compact (completely continuous) if the function defined by $p(x) = \frac{(x-a)(b-x)}{\alpha(x)}$ is integrable over $I = (a,b)$. This result can be proved with the help of the theorem of Fréchet-Kolmogorov. For a proof see MARTINI and BOER [2].

Having discussed these Hilbert spaces we have arrived at

5. INTEGRATION OF A DIFFERENTIAL EQUATION.

Consider the differential expression

$$A_\lambda = \lambda - \alpha D^2 - \beta D$$

on I. λ is a complex number. Now suppose that $u,g \in C_0^\infty(I)$ are related by

$$A_\lambda u = g.$$

Evidently this relation is equivalent to: For each $v \in C_0^\infty(I)$

(5.1) $$(A_\lambda u,v)_H = (g,v)_H$$

holds. By writing this out and by integration by parts we obtain

(5.2) $$\lambda \int_I u\bar{v}\,\frac{dx}{\alpha} + \int_I Du\,\overline{Dv}\,dx - \int_I \beta(Du)\bar{v}\,\frac{dx}{\alpha} = \int_I g\bar{v}\,\frac{dx}{\alpha}\,.$$

Let $\hat{V}$ be the linear manifold $C_o^\infty(I)$ provided with the relative topology of V. Then if we introduce the sequi-linear form $\hat{B}_\lambda$ on $\hat{V} \times \hat{V}$ by

(5.3) $$\hat{B}_\lambda(u,v) = \lambda \int_I u\bar{v}\,\frac{dx}{\alpha} + \int_I Du\,\overline{Dv}\,dx - \int_I \beta(Du)\bar{v}\,\frac{dx}{\alpha}$$

we can write for (5.2)

$$\hat{B}_\lambda(u,v) = (g,v)_H.$$

Now it can be proved that $\hat{B}_\lambda$ is a bounded sesqui-linear form on $\hat{V} \times \hat{V}$. Hence remembering that $C_o^\infty(I)$ is dense in V, $\hat{B}_\lambda$ admits a unique bounded linear extension B_λ to V × V. This sesqui-linear form can be proved to be coercive for Re λ sufficiently large; i.e., there exists a $\lambda_o > 0$ and a $c > 0$ such that

$$\operatorname{Re} B_\lambda(u,u) \geqq c\,||u||_V^2$$

Hence by the theorem of Lax-Milgram there exist a family of linear maps $\mathcal{L}_\lambda(\lambda \in C), \mathcal{L}_\lambda : V \to V$, such that

$$B_\lambda(u,v) = (\mathcal{L}_\lambda u, v)_V$$

for all u,v ∈ V and $\mathcal{L}_\lambda$ is a linear isomorphism for $\operatorname{Re}\lambda \geqq \lambda_o$.

Let J* be the adjoint of the inclusion map J: V → H. Remembering that J has dense range it follows that J* is injective. Then the following facts can be established:

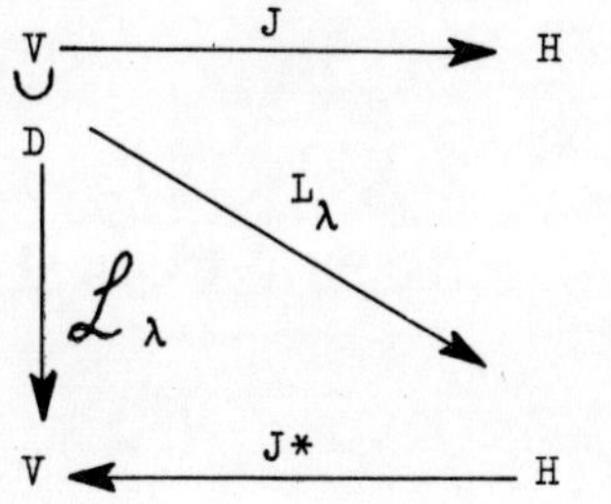

a) For $\lambda \geqq \lambda_0$ $L_\lambda = (J^*)^{-1} \mathcal{L}_\lambda$ is an injective map from the set

$$D = \{f \in V | A_0 f \in H\}$$

onto H.

b) $L_\lambda u = A_\lambda u$ for all $u \in D$.

Hence for $\lambda \geqq \lambda_0$ $A_\lambda : D \to H$ is a bijection.
For the proofs of the foregoing statements see MARTINI [1], section 5.

6. END OF THE PROOF OF THEOREM 2.

Step 4. Now let $\lambda \geqq \lambda_0$ and let λ be fixed. Denote by

$$D^*_\lambda = \{f \in D | A_\lambda f \in C_0^\infty(I)\}.$$

Then it follows that $D^*_\lambda \subset D(A)$. Let M be the linear manifold spanned by the two elements f_1, f_2 given by

$$f_1(t) = \lambda t - \beta(t) , \quad f_2(t) = 1.$$

Then we can split $C(\bar{I})$ up into the direct sum

$$C(\bar{I}) = M \oplus C_v(\bar{I}) ,$$

where $C_v(\bar{I})$ is the space of all $f \in C(\bar{I})$ which are zero at the boundary of I. We remark that $C_0^\infty(I)$ is a dense subset of $C_v(\bar{I})$. Since M is the image under λ-A of the polynomials of degree at most one and $C_0^\infty(I)$ is contained in the range of λ-A it follows by the linearity of the operator λ-A that the range of λ-A is dense in $C(\bar{I})$. This completes the proof of theorem 2.

7. A METHOD FOR CONSTRUCTION.

Abouth the construction of the semi-groups we should like to say the following:

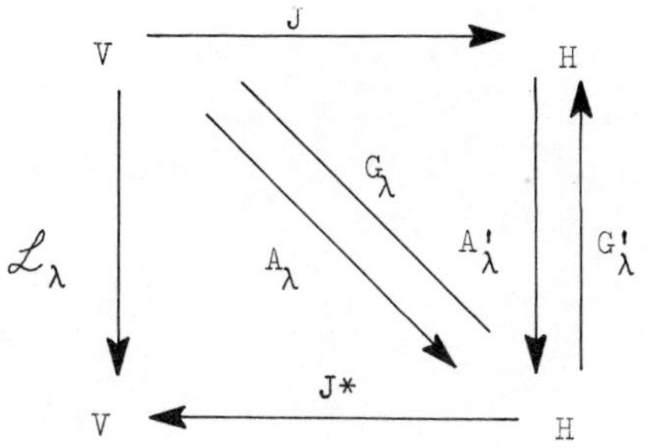

Suppose that J is compact (we know that this is the case when $p(x) = \frac{(x-a)(b-x)}{\alpha(x)}$ is integrable over $I = (a,b)$ and that A is symmetric on $C_0^\infty(I)$. Then G'_λ is a compact hermitian operator since it is the composition of a compact map and a bounded linear map. Thus we may apply the spectral theory for compact hermitian operators to G'_λ and then by a small modification of G_λ we get a representation for the resolvent $R(\lambda;A)$ and since the resolvent of the semi-group is related to the semi-group by the Laplace transform

$$R(\lambda,A) = \int_0^\infty e^{-\lambda t} \, T_t f \, dt$$

we get by inversion a representation for the semi-group $\{T_t\}(t \geqq 0)$.

For more details abouth this method and an illustration by the differential expression $A(x,D) = cx(1-x)D^2$, $c > 0$, see MARTINI and BOER [2].

REFERENCES

1. MARTINI, R., A relation between semi-groups and sequences of approximation operators.

2. MARTINI, R. and W.L. BOER, On the construction of semi-groups of operators.

3. YOSIDA, K., Functional Analysis, Springer-Verlag (1965).

DEGENERATE ELLIPTIC OPERATORS IN UNBOUNDED DOMAINS

W.D.EVANS
Pure Mathematics Department,
University College, Cardiff, Wales.

Abstract: In this article we present results concerning the essential spectrum of a self-adjoint operator T defined by a general second-order degenerate elliptic differential expression τ in an unbounded open subset Ω of $\mathbb{R}^n (n \geq 1)$. We investigate the dependence of the essential spectrum of T on the coefficients of τ and, in particular, on Ω. Also, we discuss the effect on the essential spectrum of changes inside a subset Ω' of Ω.

1. INTRODUCTION

Let Ω be an unbounded domain in $\mathbb{R}^n$ $(n \geq 1)$ and let τ be the second-order differential expression defined by

$$\tau u(x) = \sum_{j,k=1}^{n} D_j \left(a_{jk}(x) D_k u(x) \right) + q(x) u(x)$$

where $D_j u(x) = \left(\frac{\partial}{\partial x_j} + b_j(x) \right) u(x)$. The coefficients of τ are assumed to be real and to satisfy the following conditions throughout:-

I
- i) $a_{jk}, b_j \in C^1(\Omega)$, $j,k = 1,2,\cdots,n$,
- ii) $\sum_{j,k=1}^{n} a_{jk} b_j b_k$ is locally bounded on $\overline{\Omega}$,
- iii) $q \in L^1_{loc}(\Omega)$,
- iv) the matrix $\{a_{jk}(x)\}$ is symmetric and there exists a non-negative locally integrable function m such that
$$\sum_{j,k=1}^{n} a_{jk}(x) \xi_k \overline{\xi_j} \geq m(x) |\xi|^2$$
for a.e. $x \in \Omega$ and all $\xi = (\xi_1, \cdots, \xi_n)$ in the n-dimensional complex space $\mathbb{C}^n$ with $|\xi|^2 = \sum_{i=1}^{n} |\xi_i|^2$.

Hence, in view of I(iv) τ is a degenerate elliptic operator.

The self-adjoint operator T which forms the subject of the investigation will be defined in terms of the symmetric sesquilinear form

$$h(u,v) = \sum_{j,k=1}^{n} (a_{jk} D_k u, D_j v) + (qu, v) \tag{1}$$

where $(\cdot,\cdot)$ denotes the usual inner product on $L^2(\Omega)$. Our first task will be to prove that, under suitable conditions on the coefficients of τ, h is a closed densely defined form in $L^2(\Omega)$ which is bounded below. We may then invoke the well-known representation theorem for such forms to produce the desired operator T. This operator T has the property of being bounded below, which is crucial for the ensuing discussion and is in fact the reason why T appears to be the most natural operator to consider. Notice that if the minimal operator T_0 defined by τ on $C_0^\infty(\Omega)$ exists then T is its Friedrichs extension. However, as we are only supposing that $q \in L^1_{loc}(\Omega)$, the existence of T_0 is not guaranteed in general.

2. THE OPERATOR T

Before proceeding we need to introduce some notation.

We first define, for a positive real number α,

$$W_\alpha(x) = \begin{cases} |x|^{\alpha-n} & \text{if } \alpha < n, \\ 1 & \text{if } \alpha > n, \\ 1 - \log|x| \text{ when } |x| < 1 \text{ and } 1 \text{ for } |x| \geq 1 & \text{if } \alpha = n. \end{cases}$$

For a measurable function Q on Ω and $d > 0$, we set

$$M_{\alpha,d}(|Q|, x) = \int_{\Omega \cap B(x,d)} |Q(y)| \, W_\alpha(x-y) \, dy,$$

$$M_{\alpha,d}(|Q|) = \sup_{x \in \Omega} M_{\alpha,d}(|Q|, x),$$

where $B(x,d)$ is the ball centre x and radius d in $\mathbb{R}^n$. We also put

$$N_{t,d}(|Q|, x) = \left\{ \frac{1}{|B(x,d)|} \int_{\Omega \cap B(x,d)} |Q(y)|^t \, dy \right\}^{1/t},$$

$$N_{t,d}(|Q|) = \sup_{x \in \Omega} N_{t,d}(|Q|, x),$$

where $|B(x,d)|$ denotes the volume in $\mathbb{R}^n$.

Let h_0 denote the symmetric sesquilinear form

$$h_0(u,v) = \sum_{j,k=1}^{n} (a_{jk} D_k u, D_j v)$$

whose domain $\mathcal{D}_0$ in $L^2(\Omega)$ is the closure of $C_0^1(\Omega)$ with respect to the norm

$$\|u\|_{h_0} = \{ h_0[u] + \|u\|^2 \}^{1/2} ,$$

where we have written $h_0[u]$ for $h_0(u,u)$ and $\| \ \|$ is the $L^2(\Omega)$ norm. $\mathcal{D}_0$ is a Hilbert space with inner product $h_0(u,v) + (u,v)$ and the natural injection $\mathcal{D}_0 \to L^2(\Omega)$ is continuous.

The following lemmas are required.

Lemma 1 Let $N_{t,d}(m^{-1}) < \infty$ for some $t > \max(1, n/2)$ and $M_{\alpha,d}(|Q|) < \infty$ for some α satisfying $\max(-n+2, 0) < \alpha < 2 - n/t$. Let

$$c(\alpha,t) = \left\{ \omega_n^{-1} n^{-1/t} \left(\frac{1 - 1/t}{2 - \alpha - n/t} \right)^{1 - 1/t} \right\}^{1/2} ,$$

where ω_n is the $(n-1)$-dimensional measure of the unit sphere in $\mathbb{R}^n$, and put

$$\eta = \sup_{x \in \Omega} \frac{|\Omega \cap B(x,d)|}{|B(x,d)|}$$

Then $u \mapsto |Q|^{1/2} u$ is a bounded map of $\mathcal{D}_0$ into $L^2(\Omega)$ and for $u \in \mathcal{D}_0$

$$\begin{aligned} \| |Q|^{1/2} u \| &= (|Q|u, u)^{1/2} \\ &\leq (d^n W_\alpha(d))^{-1/2} M_{\alpha,d}^{1/2}(|Q|) \Big\{ c(\alpha,\infty)\, \eta^{(2-\alpha)/2n} \|u\| + \\ &\quad + c(\alpha,t)\, \eta^{(2-\alpha-n/t)/2n} N_{t,d}^{1/2}(m^{-1})\, d\, h_0^{1/2}[u] \Big\} \end{aligned}$$

Lemma 2 Let $\mathcal{M}$ be the function defined by

$$\mathcal{M}(x-a) = -r^{n-1} \int_{r_0}^{r} \frac{d\rho}{m(a+\rho\xi)\,\rho^{n-1}}$$

where $x - a = \rho\xi$, $|\xi| = 1$, are polar coordinates and suppose that r_0 can be chosen such that the integral is convergent for $r > 0$ and infinite when $r = 0$. Then for all $\varphi \in \mathcal{D}_0$

$$\int_\Omega \frac{|\varphi(x)|^2}{\mathcal{M}^2(x-a)\, m(x)}\, dx \leq 4\, h_0[\varphi] .$$

When $m = 1$ and $n = 3$ the latter inequality is the familiar one

$$\int \frac{|\varphi(x)|^2}{|x-a|^2}\, dx \leq 4\, \|\nabla\varphi\|^2$$

for $\varphi \in C_0^1(\Omega)$.

The proofs of the above lemmas and all subsequent results in this paper may be found in (Evans).

The additional assumptions which are required for establishing the existence of the operator T mentioned in §1 are determined by Lemmas 1 and 2 and are as follows:

II
(a) $N_{t,d}(m^{-1}) < \infty$ for some $t > \max(1, n/2)$,

(b) $q = q_1 + q_2 + q_3$ where

i) $q_1(x) \geq 0$ and $M_{\alpha_1,d}(q_1,x)$ is locally bounded in $\bar{\Omega}$ for some α_1 satisfying $0 < \alpha_1 < 2 - n/t$,

ii) $M_{\alpha_2,d}(|q_2|) < \infty$ for some α_2, $0 < \alpha_2 < 2 - n/t$,

iii) $q_3(x) = \sum c_j \{m^2(x-a_j)m(x)\}^{-1}$, where the summation is finite, but arbitrary, and $\sum_{c_j<0} |c_j| < 1/4$.

Under these assumptions it can be shown by a modification of standard arguments (c.f. Theorem VI-4.6 in Kato, 1966) that the symmetric sesquilinear form h in (1) has domain

$$\mathcal{D}(h) = \{u \mid u \in \mathcal{D}_0, |(q_1 u, u)| < \infty\}$$

and is closed, densely defined and bounded below in $L^2(\Omega)$. It then follows that there exists a self-adjoint operator T in $L^2(\Omega)$ which is such that its domain $\mathcal{D}(T)$ is a core of h and

$$(Tu, v) = h(u,v) \quad , u \in \mathcal{D}(T), v \in \mathcal{D}(h).$$

If further we suppose that there exists in Ω a function ρ having the properties:-

III
i) $\rho(x) \geq 0$ and $\rho(x) \to \infty$ as $|x| \to \infty$,

ii) ρ is uniformly Lipshitz on compact subsets of $\bar{\Omega}$,

iii) $\sum_{j,k=1}^{n} a_{jk}(x) \frac{\partial \rho}{\partial x_k} \frac{\partial \rho}{\partial x_j} \leq 1$ a.e on Ω,

then it can be shown that $C_0^\infty(\Omega)$ is a core of h and

$$\mathcal{D}(T) = \{u \mid u \in \mathcal{D}(h), \tau u \text{ exists in the generalised sense and } \tau u \in L^2(\Omega)\}$$

with $Tu = \tau u$. Furthermore, there exist positive constants c_1, c_2 such that

$$h[u] \geq c_1 h_0[u] - c_2 \|u\|^2$$

for $u \in \mathcal{D}(h)$ and in particular T is bounded below.

3. RELATIVELY COMPACT PERTURBATIONS

We have seen that the sesquilinear form h is bounded below in $L^2(\Omega)$ and so, for a sufficiently large positive number μ, $h+\mu$ is positive definite. We therefore have defined on $\mathcal{D}(h)$ the inner product

$$(u,v)_h = (h+\mu)(u,v) = h(u,v) + \mu(u,v),$$

and since h is closed, $\mathcal{D}(h)$ with this inner product is a Hilbert space,

$\mathfrak{D}$ say. All the norms $\|\cdot\|_h$ are equivalent for all μ for which $h+\mu$ is positive definite.

Let s be a symmetric sesquilinear form which is bounded on $\mathfrak{D}$ so that there exists a bounded self-adjoint operator S on $\mathfrak{D}$ such that

$$s(u,v) \quad = \quad (Su,v)_h .$$

Definition We say that s is h-compact if S is compact on $\mathfrak{D}$. This notion of relative compactness is similar to that in Chapter 1, § 4 of (Glazman, 1965) except that it is phrased in terms of forms. The properties of h-compact forms which we need for our discussion follow in much the same way as the analagous results proved in (Glazman, 1965).

In the applications below $s(u,v) = (Qu,v)$ where Q is a real function, $Q(x) = h(x) - q(x)$, which is assumed to satisfy Lemma 1. In Theorem 3 below we obtain conditions for the map $u \mapsto |Q|^{1/2} u$ to be a compact map of $\mathfrak{D}_0$ into $L^2(\Omega)$. From this it follows that s is h-compact and this in turn implies that the sesquilinear form

$$h'(u,v) \quad = \quad h(u,v) \quad + \quad s(u,v) \qquad , \qquad u,v \in \mathfrak{D}(h)$$

is closed in $L^2(\Omega)$ and bounded below. It therefore defines a self-adjoint operator T' which is bounded below and a further consequence of the h-compactness of s is that T and T' have the same essential spectrum. By the essential spectrum $\sigma_E(T)$ of T we mean the complement in the spectrum of T of the isolated eigenvalues of finite multiplicity.

Since $Q(x) = h(x) - q(x)$ and h satisfies the same conditions as q (see assumptions II), the operator T' above can be described in the same way as T in § 2 with q replaced by h. Note that we can not assert that $T' = T + Q$ since Q is not defined on the domain of T. However, this is unnecessary for our purpose.

In order to indicate the dependence of T on q we write T_q. We also need to define sets of the form

$$K(R) \quad = \quad \{ x \in \Omega \mid \rho(x) \le R \}$$

where ρ is the Lipshitz function on Ω defined in III.

Theorem 3 Let $N_{t,d}(m^{-1}) < \infty$ for some $t > \max(1, n/2)$ and let $M_{\alpha,d}(|Q|) < \infty$ for some α satisfying $0 < \alpha < 2 - n/t$. Suppose also that any one of the following conditions holds:-

i) $$\lim_{\substack{|x| \to \infty \\ x \in \Omega}} \frac{|\Omega \cap B(x,d)|}{|B(x,d)|} = 0 ,$$

ii) $\liminf_{R\to\infty} M_{\alpha,d}(|Q_{\Omega\setminus K(R)}|) = 0$,

where
$$Q_{\Omega\setminus K(R)}(x) = \begin{cases} Q(x) & \text{in } \Omega\setminus K(R), \\ 0 & \text{otherwise}. \end{cases}$$

Then $u \mapsto |Q|^{1/2}u$ is a compact map of $\mathcal{D}_0$ into $L^2(\Omega)$.

When $Q(x)=1$, Q is the embedding map $\mathcal{D}_0 \subset L^2(\Omega)$ and so if (i) holds this embedding is compact. In the special case $m=1$, $a_{jk}=\delta_{jk}$, $\mathcal{D}_0$ is the Sobolev space $H_0^{1,2}(\Omega)$ and we therefore have a sufficient condition for the embedding $H_0^{1,2}(\Omega) \subset L^2(\Omega)$ to be compact. The compactness of such embeddings for unbounded domains Ω has recently attracted a great deal of interest and in (Adams, 1970) a necessary and sufficient condition on Ω for such embeddings to be compact is given. The condition (i) is not necessary for compactness of the embedding, despite the fact that if Ω contains a sequence of balls of some fixed (but arbitrary) radius then the embedding is not compact. For, it is known that the embedding is compact if Ω is the 'spiny urchin' (see Clark, 1967) but in this case (i) is not satisfied.

As mentioned above, it follows from Theorem 3 that (Qu,v) is h-compact and that $\sigma_E(T_h) = \sigma_E(T_q)$. An immediate corollary is

Corollary 4 Let Ω' be a subset of Ω . Let $h(x)=q(x)$ in $\Omega\setminus\Omega'$ and suppose that $Q(x) = h(x) - q(x)$ is such that $M_{\alpha,d}(|Q|) < \infty$ for some α , $0<\alpha<2-n/t$ $(t>\max(1,n/2))$. Then if

$$\liminf_{R\to\infty} M_{\alpha,d}(|Q_{\Omega\setminus K(R)}|) = 0$$

we have that $\sigma_E(T_h) = \sigma_E(T_q)$.

If Ω' is bounded in Corollary 4 we get the decomposition principle of Birman and Glazman (see §§ 17,20 in Glazman 1965). It means that $\sigma_E(T_q)$ depends only on the behaviour of q at infinity. A result similar to Corollary 4 but for non-degenerate τ was obtained by V.Beck in an unpublished thesis (see Theorem 9 in Jorgens, 1970).

4. DEPENDENCE OF σ_E ON Ω

In order to establish the main results on the essential spectrum we require the following additional assumption:

IV $\begin{cases} \text{The negative part } q^- \text{ of } q \text{ is such that } M_{\alpha,d}(|q^-|,x) \text{ is locally bounded} \\ \text{for some } \alpha,\ 0<\alpha<2-n/t. \end{cases}$

This is satisfied if $q_3^-=0$ in II.

Theorem 5 Suppose that

$$\liminf_{\substack{|x|\to\infty \\ x\in\Omega}} q(x) = \beta .$$

Let

$$\eta = \limsup_{\substack{|x|\to\infty \\ x\in\Omega}} \frac{|\Omega \cap B(x,d)|}{|B(x,d)|} \tag{2}$$

and, for $\eta > 0$ (for all $d > 0$) and t as in II(a), put

$$\ell(\eta) = \sup_{d>0} \left\{ \left(\frac{1-\eta^{1/n}}{\eta^{1/n}}\right)^2 \frac{(n\eta)^{1/t}}{d^2 N_{t,d}(m^{-1})} \left(\frac{2-n/t}{2-1/t}\right)^{2-1/t} \right\}. \tag{3}$$

Then, if $\eta > 0$,

$$\sigma_E(T_q) \subseteq [\beta + \ell(\eta), \infty).$$

If $\eta = 0$ for some (and hence all) $d > 0$ then $\sigma_E(T_q) = \phi$ so that T_q has a purely discrete spectrum.

The estimated lower bound for $\sigma_E(T_q)$ in Theorem 5 would be at its most precise when Ω approximates a disjoint union of balls in $\mathbb{R}^n$. For other domains the same method would yield a better estimate if the balls $B(x,d)$ were replaced by other units. For instance if Ω is a strip in $\mathbb{R}^2$, squares would be most suitable. When Ω is a strip in $\mathbb{R}^2$ of width 2ρ and $\tau = -\Delta + q$, Theorem 5 gives $\ell(\eta) \simeq \cdot 04\rho^{-2}$, whereas when $q(x) = \beta$ it can easily be shown by separation of variables that $\beta + \frac{\pi^2}{4}\rho^{-2}$ is the lower bound of the essential spectrum. However less uniform domains are not amenable to such direct methods.

The other question we wish to investigate is the following. Let Ω' be an unbounded subset of Ω and suppose that

$$\liminf_{\substack{|x|\to\infty \\ x\in\Omega}} q(x) = \beta , \tag{4}$$

$$\liminf_{\substack{|x|\to\infty \\ x\in\Omega\setminus\Omega'}} q(x) = \gamma > \beta \tag{5}$$

We have from Theorem 5 that $\sigma_E(T_q) \subseteq [\beta + \ell(\eta), \infty)$. But, when can it be said that $\sigma_E(T_q) \subseteq [\delta + \ell(\eta), \infty)$ for $\delta = \gamma$ and also for $\beta < \delta < \gamma$? This question was answered in (Eastham, 1967) for the case $\Omega = \mathbb{R}^n$ and $\tau = -\Delta + q$. Eastham obtained his results by the use of various comparison techniques and perturbation theory and these methods do not appear to extend to the general problem. For the solution of the general problem given in Theorem 6 below we rely heavily on Lemma 1 and Theorem 3, particularly the special case of $Q(x) = 1$ when Q is the embedding map of $\mathcal{D}_0$ into $L^2(\Omega)$.

Theorem 6 Let $\eta\ (>0)$, $\ell(\eta)$, β and γ be as in (2),(3),(4) and (5) and let

$$\eta' = \limsup_{\substack{|x|\to\infty \\ x\in\Omega}} \frac{|\Omega' \cap B(x,d)|}{|B(x,d)|} .$$

We then have the following results.

a) If $\eta' = 0$ for some (and hence all) $d > 0$ then $\sigma_E(T_q) \subseteq [\gamma + \ell(\eta), \infty)$.

b) If $\eta = 1$ for all $d > 0$ (so that $\ell(\eta) = 0$) and $0 < \eta' < 1$ for some $d > 0$ then there exists a κ, $0 < \kappa < (\gamma - \beta)(1 - \eta'^{1/n})$ such that $\sigma_E(T_q) \subseteq [\beta + \kappa, \infty)$.

Case (a) can be regarded as an extension of the decomposition principle to unbounded domains Ω' which are 'arbitrarily thin' at infinity. The condition on Ω' in (b) implies that the embedding map $H_0^{1,2}(\Omega) \subset L^2(\Omega)$ is a k-set contraction for some $k \leq \eta'^{1/n} < 1$ (see Edmunds, Corollary 3.14).

REFERENCES

Adams, R.A., 1970, Capacity and compact imbeddings, J.Math.Mech. 19, 923.

Clark, C.W., 1967, Rellich's embedding theorem for a 'spiny urchin,' Canad.Math.Bull. 10, 731.

Eastham, M.S.P., 1967, On the discreteness of the spectrum in eigenfunction theory, J.London Math.Soc. 42, 309.

Edmunds, D.E. and Evans, W.D., to appear, Elliptic and degenerate elliptic operators in unbounded domains.

Evans, W.D., to appear, On the essential spectrum of second order degenerate elliptic operators.

Glazman, I.M., 1965, Direct methods of qualitative spectral analysis of singular differential operators (Israel Program for Scientific Translations, Jerusalem).

Jorgens, K., 1970, Spectral theory of Schrödinger operators, (Lectures delivered at the University of Colorado).

Kato, T., 1966, Perturbation theory of linear operators, (Springer).

SCATTERING THEORY FOR A GENERAL CLASS OF DIFFERENTIAL OPERATORS

KRESIMIR VESELIĆ
Institut Ruđer Bošković, Zagreb
and
JOACHIM WEIDMANN
Fachbereich Mathematik der Universität Frankfurt am Main

ABSTRACT: In this paper we give an account of some recent results on the existence of wave operators for very general unperturbed operators and perturbations. In section 5 we give a completeness result under similar conditions for the case of dimension 1.

1. INTRODUCTION

In this section we will explain some important notions from mathematical scattering theory, see Kato (1966) Chapter X.

Let T_o and T be self-adjoint operators in a Hilbert space H; T_o is called the unperturbed operator, T the perturbed operator and $V := T-T_o$ the perturbation. In scattering theory one considers the wave operators

$$W_\pm = W_\pm(T,T_o) := \operatorname*{s-lim}_{t\to\pm\infty} e^{itT} e^{-itT_o} P_o,$$

if these limits exist; here P_o is the orthogonal projection onto the absolutely continuous subspace H_{oa} of T_o, i.e. the space of $u \in H$ for which $\lambda \mapsto \langle u, E_o(\lambda)u\rangle$ is absolutely continuous, where E_o is the spectral resolution of T_o. In most cases we have $P_o = I$; this is so in all cases considered below. Anyhow, there are reasons to put the P_o there; for example, if we want to consider $W_\pm(T_o,T)$, then the projection P onto the absolutely continuous subspace H_a of T is needed.

The operators $W_\pm$ (if they exist) are partial isometries with initial set H_{oa} and final set $R(W_\pm) \subset H_a$. $T_o|_{H_{oa}}$ is then unitarily equivalent to $T|_{R(W_+)}$ and $T|_{R(W_-)}$, where the unitary equivalence is given by W_+ and W_- respectively. We also have the intertwining property

$$W_+T_o|_{H_{oa}} = T|_{R(W_+)}W_+, \quad W_-T_o|_{H_{oa}} = T|_{R(W_-)}W_- .$$

These notions are motivated by physical considerations. The aim is to define the scattering operator

$$S := W_+^* W_-$$

which is expected to be unitary as an operator in H_{oa} (it was shown by Kato-Kuroda (1959) that this does not hold automatically, if the wave operators exist). Since S is unitary if and only if $R(W_+) = R(W_-)$, it is natural to pose the following problems

(I) Existence of wave operators. Give conditions on T_o and T which guarantee the existence of wave operators.

(II) Completeness of wave operators. Give conditions on T_o and T such that (in addition to (I))

$$R(W_+) = R(W_-) = H_a .$$

There is a simple condition which answers problem (II): The wave operators $W_\pm(T,T_o)$ are complete if and only if the wave operators $W_\pm(T_o,T)$ exist, see Kuroda (1959). But this condition is not easy to be verified.

2. SOME KNOWN RESULTS

It is the purpose of this section to recall some typical results. Many other results have been published recently, but we cannot mention all of them.

The existence theorems for wave operators are usually proved by means of the following abstract theorem which is due to Cook.

2.1. Theorem [see Kato (1966) X.3.7]. Let D be a dense subspace of H_{oa} such that for every $u \in D$

$$e^{-itT_o}u \in D \quad \text{for every } t \in \mathbb{R},$$

$$\mathbb{R} \ni t \mapsto (T-T_o)e^{-itT_o}u \in H$$

is continuous and

$$\int_{-\infty}^{\infty} \|(T-T_o)e^{-itT_o}u\|dt < \infty.$$

Then the wave operators $W_\pm$ exist. (Recently Eckardt (1973) has shown that the continuity assumption is not needed).

The following abstract result is due to Birman, Kato and Rosenblum; it guarantees the existence and the completeness of the wave operators at the same time, and it also contains the so called invariance principle.

Let us first define by Φ the class of functions φ: $\mathbb{R} \to \mathbb{R}$ with the following properties: $\mathbb{R}$ can be devided into a finite number of subintervals such that in the interior of every subinterval φ is differentiable, φ' is continuous, locally of bounded variation and of constant sign.

2.2. Theorem [see Kato (1966) X.4.7]. Let $V := T - T_o$ be in the trace class. Then for every $\varphi \in \Phi$ the wave operators $W_\pm(\varphi(T),\varphi(T_o))$ exist and are complete. If I^+ and I^- are the subsets of $\mathbb{R}$ on which φ is increasing or decreasing, respectively, we have the invariance principle

$$W_\pm(\varphi(T),\varphi(T_o))E(I^+) = W_\pm(T,T_o)E(I^+)$$

$$W_\pm(\varphi(T),\varphi(T_o))E(I^-) = W_\mp(T,T_o)E(I^-).$$

We now turn to one of the most important cases: $T_o = -\Delta$ in $L_2(\mathbb{R}^m)$ and $V := T - T_o$ is a multiplication operator. Besides some recent results on scattering theory for more general elliptic operators, most results are concerned with this case. We state two results.

2.3. Existence [Kuroda (1959)]. The existence of $W_\pm(T,T_o)$ is guaranteed if

$$V(.)(1+|.|)^{1-\frac{m}{2}+\varepsilon} \in L_2(\mathbb{R}^m) \text{ for some } \varepsilon > 0.$$

2.4. Completeness [Kato (1969)]. The existence and completeness of $W_\pm(T,T_o)$ is guaranteed if

$$V(.)(1+|.|)^{-\Theta} \text{ is bounded for some } \Theta < -1.$$

(The existence follows in this case from 2.3 and is actually due to Hack).

In sections 4 and 5 we intend to generalize these results essentially in two directions:

(a) T_o belongs to a general class of (pseudo-) differential operators. (Also operators in $[L_2(\mathbb{R}^m)]^M$ can be considered, see Veselić-Weidmann (1973 a).)

(b) there is no restriction on the nature of the operator V; essentially we use a smallness condition near infinity (theorems 4.1, 4.3 and 5.1) or a weighted L_p-estimate (Theorem 4.5).

We shall mainly study the existence problem (I); in section 5 we shall indicate how completeness results may be proved by means of theorem 2.2.

3. MOTIVATION FOR OUR CONDITIONS

In Jörgens-Weidmann (1973 a) it became apparent, that for the invariance of the essential spectrum σ_e of a Hamiltonian operator under perturbations it is not needed, that this perturbation is relatively compact, but only that it is relatively small for functions which have their support near infinity.

3.1. Definition. Let T and V be operators in $L_2(\mathbb{R}^m)$. V is called T-small at infinity, if V is T-bounded and for every $\varepsilon > 0$ there is an $r(\varepsilon) \geq 0$ such that

$$\|Vu\|_2 \leq \varepsilon \, (\|u\|_2 + \|Tu\|_2)$$

for every $u \in D(T)$ with $u(x) = 0$ for $|x| \leq r(\varepsilon)$.

3.2. Remark [Jörgens-Weidmann (1973 a),3.12]. If T is a Schrödinger operator in the sense of Jörgens-Weidmann (1973 a) (for example if $T = -\Delta$), then we have: V is T-compact if and only if V is T-small at infinity and has T-bound zero.

The following result holds.

3.3. Theorem [Jörgens-Weidmann (1973a) 3.19 and Böcker (1973) Theorem 1].
Let T_o be a Schrödinger operator in the sense of Jörgens-Weidmann (1973a) and V a symmetric operator which is T_o-small at infinity such that $V+T_o$ is self-adjoint. Then we have $\sigma_e(T_o+V) = \sigma_e(T_o)$.

Comparing this result with other results on the invariance of the essential spectrum and with existence results for wave operators Jörgens and one of the authors were led to the conjecture that an existence result for general perturbations V should hold if the decrease of the operator V near infinity is faster than $\frac{1}{r}$.

4. NEW EXISTENCE RESULTS

The first result in this direction was

4.1. Theorem [Jörgens-Weidmann (1973b)]. Let $T_o = -\Delta$, V a symmetric operator with $S(\mathbb{R}^m) \subset D(V)$ such that there exist self-adjoint extensions of $T_1 := T_o+V$ and constants $C > 0$ and $\Theta < -1$ such that for every $r \geq 0$

$$\|Vu\|_2 \leq C(1+r)^{\Theta} \{\|u\|_2 + \|T_o u\|_2\}$$

for every $u \in D(T_o)$ with $u(x) = 0$ for $|x| \leq r$.
Then the wave operators $W_{\pm}(T,T_o)$ exist for every self-adjoint extension T of T_1.

We indicate the proof by using theorem 2.1 and a method which has been introduced by Veselić-Weidmann (1973a) in the more general situation given below.

Proof. For the dense subspace D in theorem 2.1 we use the functions $u \in S(\mathbb{R}^m)$ such that $\hat{u} \in C_o^{\infty}(\mathbb{R}^m \setminus \{y \in \mathbb{R}^m : y_1 = 0\})$ ($\hat{u}$ is the Fourier transform of u). Then for

$$u(t,x) = (e^{-itT_o} u)(x)$$

and every differential operator

$$D = p(\partial), \text{ where } p \text{ is a polynomial in } m \text{ variables}$$

we have (D operates with respect to x; C_k depends also on u and D but not on x and t)

$$|Du(t,x)| \leq C_k(1+|x|)^k |t|^{-k}, \quad k \in \mathbb{N}_o = \{0\} \cup \mathbb{N}.$$

We prove this for $k = 1$:

$$Du(t,x) = (2\pi)^{-m/2} \int e^{ixy-ity^2} p(iy)\hat{u}(y)dy$$

$$(4.2) \quad = -(2\pi)^{-m/2} \int [e^{-ity^2}(-2ity_1)] \left[\frac{1}{2ity_1} p(iy)e^{ixy}\hat{u}(y)\right]dy$$

$$= (2\pi)^{-m/2} \frac{-i}{2t} \int e^{-ity^2} \frac{d}{dy_1} \left[\frac{p(iy)}{y_1} e^{ixy}\hat{u}(y)\right]dy.$$

By means of the choice of u this implies the result for $k = 1$. For $k > 1$ the result follows by an inductive procedure.

We now cut $u(t,x)$ in the x-space smoothly into two parts ($\alpha > 0$)

$$u_1 \quad \text{for} \quad |x| \leq |t|^{\alpha}+1$$

$$u_2 \quad \text{for} \quad |x| > |t|^{\alpha}.$$

From our assumption on V and the above estimate it can now be deduced that for $|t| > 1$

$$\|Vu_1\|_2 \leq c_1(\|u_1\|_2 + \|\Delta u_1\|_2)$$
$$\leq c_2|t|^{\alpha\frac{m}{2}}\left(1+|t|^{\alpha}\right)^k |t|^{-k} \leq c_3|t|^{k(\alpha-1)+\frac{\alpha m}{2}},$$
$$\|Vu_2\|_2 \leq c_4\left(1+|t|^{\alpha}\right)^{\Theta}\left(\|u\|_2 + \|\Delta u_2\|\right) \leq c_5|t|^{\alpha\,\Theta}.$$

If we choose α such that $\frac{1}{|\Theta|} < \alpha < 1$, then there exists $k \in \mathbb{N}$ such that

$$k(\alpha-1) + \frac{\alpha m}{2} < -1 \quad \text{and} \quad \alpha\,\Theta < -1.$$

Therefore the condition of Theorem 2.1 is fulfilled. Q.E.D.

So far we did not make much use of the properties of $T_o = -\Delta$. We might therefore expect that a more general theorem holds; this is in fact true:

<u>4.3. Theorem</u> [Veselić-Weidmann (1973a), Satz 2.4]. Let h be a real valued function

$$h \in C^{\infty}(\mathbb{R}^m \setminus Z), \text{ grad } h(x) \neq 0 \text{ for } x \notin Z,$$

where Z is a closed set of measure zero. Let $T_o := F^{-1}\hat{T}_h F$, where F is the Fourier transform and $\hat{T}_h$ the self-adjoint operator of multiplication by h. Assume that V is symmetric with $S(\mathbb{R}^m) \subset D(V)$ such that $T_1 := T_o + V$ has a self-adjoint extension and that there exist $\Theta < -1$, $q \in [2,\infty]$ and a differential operator D such that for every $r \geq 0$

$$\|Vu\|_2 \leq C(1+r)^{\Theta}\|Du\|_q$$

for every $u \in S(\mathbb{R}^m)$ with $u(x) = 0$ for $|x| < r$.
Then the wave operators $W_{\pm}(T,T_o)$ exist for every self-adjoint extension T of T_1.

This result is rather general in some sense; but it does not contain, for example, Kuroda's result 2.3. This result can be recovered by using the fact that integrals of the form

$$\int e^{-ity^2}\varphi(y)dy, \ \varphi \in C_o^{\infty}(\mathbb{R}^m)$$

(such as the final term in (4.2)) can be estimated by

$$c|t|^{-m/2}$$

(see Kato (1966) IX.1.8.). Using this we get for

$T_o = -\Delta$, by means of a similar technique as in the proof of Theorem 4.1, see Veselić-Weidmann (1973b),

(4.4) $$|u(x,t)| \leq C_k(1+|x|)^k \, |t|^{-k-m/2}, \quad k \in \mathbb{R}.$$

By a purely technical calculation we get from (4.4) the existence of wave operators $W_{\pm}(T,T_o)$ if $T_o = -\Delta$, T is a self-adjoint extension of $T_o + V$ and V satisfies the conditions of theorem 4.5 which holds for more general operators T_o.

4.5. Theorem [Veselić-Weidmann (1973b), Theorem 3.1]. Let h and T_o be as in Theorem 4.3 and assume in addition

$$\mathrm{rank}[\partial_i \partial_j h(y)] \geq n \quad \text{for} \quad y \in \mathbb{R}^m \setminus Z$$

for some n. Let furthermore V be a symmetric operator with $D(V) \supset S(\mathbb{R}^m)$ such that $T_o + V$ has self-adjoint extensions. Then the wave operators $W_{\pm}(T,T_o)$ exist for any self-adjoint extension T of $T_o + V$, provided that one of the following conditions is satisfied.

a) There exist $q \in (2,\infty]$, $\Theta < \frac{n}{2}\left(1 - \frac{2}{q}\right) - 1$ and $f \in C^{\infty}(\mathbb{R}^m \setminus Z)$ such that

$$\|Vu\|_2 \leq \left\|\left(1+|\cdot|^2\right)^{\Theta/2} F^{-1}(f\hat{u})\right\|_q$$

holds for every u with $u \in C_o^{\infty}(\mathbb{R}^m \setminus Z)$.

b) There exist $q \in [2,\infty]$, $\Theta < \min\left\{0, \frac{n}{2}(1-\frac{2}{q})-1\right\}$ and a differential operator D with constant coeffizients such that for every $r \geq 0$

$$\|Vu\|_2 \leq (1+r)^{\Theta} \|Du\|_q$$

for every $u \in S(\mathbb{R}^m)$ with $u(x) = 0$ for $|x| \leq r$.

The proof of this theorem [Veselić-Weidmann (1973b)] uses the technique of the Morse Lemma which reduces the problem "locally" to $T_o = -\Delta$.

Some applications of this result are given in Veselić-Weidmann (1973b). For example Kuroda's result can be recovered from part a) with $n = m$ and $q = \infty$.

5. A COMPLETENESS RESULT

It is not difficult to show that the conditions of Theorem 4.5.b) are in general (at least for $m \geq 4$) not sufficient for the completeness of the wave operators. We shall here indicate how completeness results might be obtained for the situation considered in Theorem 4.1.

It is a simple consequence of Theorem 2.2 that for semi-bounded operators T_o and T for the existence and completeness of the wave operators it is sufficient that

$$R_o^k - R^k \in \text{trace class} \quad \text{for some } k \in \mathbb{N},$$

where $R_o = (\lambda - T_o)^{-1}$, $R = (\lambda - T)^{-1}$ and λ is smaller than the lower bounds of T_o and T. In this connection the following identity might be useful, which holds if $D(T_o) = D(T)$

$$(5.1) \quad R_o^{2^n} - R^{2^n} = R_o^{2^n} VR + R_o^{2^n-1} VR^2 + \ldots + R_o^2 VR^{2^n-1} + R_o VR^{2^n};$$

this equality follows from the second resolvent equation by a simple induction. As in Ikebe-Tayoshi (1968) we want to use this identity for n = 1, i.e.

$$R_o^2 - R^2 = R_o^2 VR + R_o VR^2 = R_o^2 VR + RVR_o R$$

$$= (R_o^2 V)R + R(VR_o^2)(\lambda - T_o)R.$$

Since $(R_o^2 V)^* = VR_o^2$ and since the operators R and $(\lambda - T_o)R$ are bounded, it is sufficient to prove that VR_o^2 is in the trace class. This we have to do in order to prove the following theorem

<u>5.2. Theorem.</u> Let $T_o = -\dfrac{d^2}{dx^2}$ in $L_2(\mathbb{R})$ and assume that V is a symmetric operator in $L_2(\mathbb{R})$ such that $D(T_o) \subset D(V)$, $T = T_o + V$ is self-adjoint and there exist constants $C > 0$ and $\Theta < -1$ such that for every $r \geq 0$

$$\|Vu\|_2 \leq C(1+r)^{\Theta}\{\|u\|_2 + \|T_o u\|_2\}$$

for every $u \in D(T_o)$ with $u(x) = 0$ for $|x| \leq r$. Then the wave operators $W_{\pm}(T, T_o)$ (which exist by Theorem 4.1) are complete.

Proof. As mentioned above it suffices to show that VR_o^2 is in the trace class. Therefore we write

$$VR_o^2 = \sum_{n \in \mathbb{Z}} V\varphi_n R_o^2$$

where $\varphi_n(x) = \varphi(x-n)$ and φ is a smooth function with support in $(-1,+1)$ such that $\sum_{n \in \mathbb{Z}} \varphi_n(x) = 1$ for every $x \in \mathbb{R}$. We then make the following considerations:

$R_o^2 : L_2(\mathbb{R}) \to D(T_o^2) = W_2^4(\mathbb{R})$ is bounded, where $W_2^4(\mathbb{R})$ is the Sobolev space of fourth order.

The multiplication by φ_n can be considered as an operator $\varphi_n : W_2^4(\mathbb{R}) \to W_2^2(n-1,n+1)$. This operator is known to be a trace class operator (see Yosida (1966)X.2). Because of the translation invariance the trace norm of φ_n is independent of n.

$V : W_2^2(n-1,n+1) \to L_2(\mathbb{R})$ is bounded and the norm of this operator is $\leq C(|n|+2)^{\Theta}$ by our assumption (remember that $W_2^2(n-1,n+1)$ is a subspace of $D(T_o) = W_2^2(\mathbb{R})$ also in the topological sense).

Putting all this together we see that $VR_o^2 = \sum_n V\varphi_n R_o^2$ is in the trace class.

This theorem solves a problem posed at the end of Jörgens-Weidmann (1973b); i.e. the absolutely continuous spcetrum of the operator T has multiplicity 2.

Theorem 5.1 can be immediately generalized to semi-bounded differential operators T_o in $L_2(\mathbb{R})$ of higher orders. Problems arise for operators in $L_2(\mathbb{R}^m)$, $m > 1$; in this case it might be useful to apply equation (5.1) for $R_o^{2n} - R^{2n}$.

REFERENCES

Böcker,U.,1973, Invarianz des wesentlichen Spektrums bei Schrödingeroperatoren. (to appear).

Eckardt,K.-J.,1973, On the existence of wave operators for Dirac operators. Manuscripta Math.(to appear).

Ikebe,T.and T.Tayoshi, 1968, Wave and scattering operators for second order elliptic operators in $\mathbb{R}^3$. Publ. of the Research Institute for Math. Sc., Kyoto University,Ser.A, 4, 483-496.

Jörgens,K. - Weidmann,J.,1973a, Spectral properties of Hamiltonian operators. Lecture Notes in Mathematics, Vol. 313, Berlin-Heidelberg-New York: Springer.

_____,1973b, Zur Existenz der Wellenoperatoren. Math.Z.131, 141-151.

Kato,T.,1966, Perturbation theory for linear operators. Berlin-Heidelberg-New York: Springer.

_____,1969, Some results on potential scattering. Proceedings of the International Conference on Functional Analysis and Related Topics, Tokyo: University of Tokyo Press.

Kato,T. and S.T.Kuroda, 1959, A remark on the unitary property of the scattering operator. Nuovo Cimento 14 , 1102-1107.

Kuroda, S.T., 1959, On the existence and unitary property of the scattering operator. Nuovo Cimento 12, 431-454.

Veselić, K. - Weidmann, J., 1973a, Existenz der Wellenoperatoren für eine allgemeine Klasse von Operatoren. (to appear in Math.Z.).

_____, 1973b, Asymptotic estimates of wave functions and the existence of wave operators. (to appear).

Yosida, K., 1966, Functional analysis. 2nd Ed. Berlin-Heidelberg-New York: Springer.

THE DOMAINS OF SELF-ADJOINT EXTENSIONS OF A SCHRÖDINGER OPERATOR

IAN M MICHAEL

Department of Mathematics, University of Dundee, Scotland, U.K.

Abstract: A basic problem in the analysis of formally self-adjoint differential expressions is to characterise the self-adjoint operators, in an appropriate Hilbert space, associated with these expressions. In a paper of 1963 (Quart. J. Math. Oxford, 14(1963), 41-45), Everitt gave connected proofs of the characterisations of the domains associated with the ordinary differential expression L, where

$$L\psi(x) = -(p(x)\psi'(x))' + q(x)\psi(x) \quad (0 \leqslant x < \infty),$$

with p,q real-valued and $p(x) > 0$ for all $x \geqslant 0$. The nature of these domains depends on whether L is limit-point or limit-circle, in the sense of Weyl, at infinity.

It is natural to look for similar characterisations for the corresponding partial differential operators. Although a complete characterisation seems unknown, Everitt's results prompt the investigation of certain domains and indicate tests to apply to particular examples.

This lecture presented the contents of the paper with the same title, by D P Goodall and I M Michael, published shortly after the conference in the Journal of the London Mathematical Society, Second Series, Volume 7 (1973), 265-271.

SPACES OF GENERALISED FUNCTIONS ASSOCIATED WITH LINEAR OPERATORS

MAGNUS GIERTZ
Department of Mathematics, Royal Institute of Technology
Stockholm, Sweden

1 INTRODUCTION

In order to extend the domains of certain differential operators, several methods have been devised to adjoin elements like δ-functions to spaces of pointwise defined functions. One way of introducing such »improper elements» is to regard them as continuous linear functionals on a space of »test functions» equipped with a suitable (metric) topology. As such, they are often referred to as *distributions*. Another way is to define them as *generalised functions* by a form of weak completion of spaces of test functions.

The space S′ of tempered distributions is an example of an extension of $L^2(-\infty, \infty)$ which may be obtained by the above two methods. Here the space S of test functions consists of all infinitely differentiable functions f on the real line which have the property that $x^m(D^n f)(x)$ tends to zero as $|x|$ tends to infinity for all natural numbers n and m. A topology on S is introduced by the distance function

$$d(f,g) = \sum_{p=0}^{\infty} \frac{\|f-g\|_p}{1+\|f-g\|_p} \quad (f,g \in S),$$

where, for each natural number p, the norm $\|f\|_p$ is defined as the maximum of $|x^m D^n f|(x)$ when $x \in (-\infty, \infty)$ and $m \leqslant p$, $n \leqslant p$. The space of all continuous linear functionals on the above metric space S provides us with one representation of the tempered distributions.

A different representation is obtained by adding to S the missing limit elements of sequences $(f_n)_{n=1}^{\infty}$ in S which have the property that, for each g in S, the sequence of inner products (f_n,g) tends to a finite limit as $n \to \infty$. On analogy with standard completion procedures, the tempered distributions now are defined as equivalence classes of such sequences.

The above two methods give the same end result in this case, as in many other cases. Still, there is a fundamental theoretical difference between them. The first method gives the distributions as a subspace of the algebraic conjugate S^f of S, that is, of the space of all linear functionals on S. With the second method they appear instead, essentially, as elements in the second algebraic conjugate S^{ff}. The distinction in this case may be somewhat obscured by the existence of a natural embedding of S in S^f, but I hope it will be made more clear shortly. It is, in a sense, the possibility of identifying test functions with continuous linear functionals on S which makes the first method work. In contrast to this, the »completion method» relies on the canonical embedding which identifies an arbitrary linear space with a subspace of its second algebraic conjugate. The resulting space of generalised functions therefore becomes, directly, an extension of the original function space, an extension consisting of objects to which we may approximate by ordinary functions.

To make what I have just said more precise, I shall begin by considering extensions of a linear space in general, in the form of subspaces of its second algebraic conjugate, which may be obtained by weak completion procedures. Each extension will be determined by a subspace of the first algebraic conjugate; the point of the whole construction is that this can be done in such a way that the extension becomes the domain of a »natural» continuation of a given linear operator, or set of operators. The concept of natural extension of an operator A is necessarily a rather vague one, in this context another vague formulation is that the properties of A should be preserved. For instance, if A is the differentiation operator restricted to some subspace of a Hilbert function space H it might seem natural that an extension of A to H should be just differentiation on all absolutely continuous functions in H. But then many different operators (determined by, say, different boundary conditions) would have the same extension. In other words, some properties of A would be forgotten.

As we shall see, the method I intend to discuss here does not extend an operator of the form »differentiation subject to boundary conditions» to differentiation. Instead, the extended operator becomes, on absolutely continuous functions, differentiation plus a term involving δ-functions which »remembers» the original boundary conditions.

In this talk I shall consider only the extension to spaces of generalised elements of operators defined on an inner product space. The method does apply to linear operators in general, provided the range of the transpose operator is not too thin, but it is only in the inner product case that we have a connexion with the »distributional» approach to generalised elements. To support my claim that the extensions in question are natural ones I then intend to give some result concerning self-adjoint operators. Finally, I shall give a list of examples, among other things to show how classical spaces of distributions may be regarded as generalised elements with respect to specific differential operators.

Most of the results discussed in this talk have been published in the Pacific Journal of Mathematics, Vol. 23, No 1, 1967 pp 47–67, where further details and references may be found.

2 EXTENSIONS OF LINEAR SPACES

Let me begin with some notation. The vector space of all linear functionals on a linear space X (the algebraic conjugate of X) is here denoted by X^f, and the value of a functional x′ at the vector x by $\langle x,x'\rangle$. A subset M of X^f is called *total* if it is only for x = 0 that $\langle x,m\rangle = 0$ for all m in M. This means that a total subset of X^f contains many enough functionals in order to distinguish between the elements of X, and ensures the existence of a natural injective imbedding of X in M^f. All subspaces of functionals considered here are assumed to have this property.

A denumerable sequence $(x_n)_{n=1}^{\infty}$ in X is called M–*regular* if $\lim_{n\to\infty} \langle x_n,m\rangle$ exists for all m in M. Each such sequence may be regarded as an M-regular sequence in M^f as well; as such it always has a limit element $\widetilde{x}$ in M^f defined by $\langle m,\widetilde{x}\rangle = \lim_{n\to\infty} \langle x_n,m\rangle \quad (m \in M)$. Corresponding to each total subset M of X^f we obtain an extension $\widetilde{X} = \widetilde{X}(M)$ of X by adjoining to this space all the missing limit elements of M-regular sequences in X. The totality of M ensures that $\widetilde{X}$ contains X, this is so since the canonical embedding of X in M^f is injective.

The extension $\widetilde{X}$ is said to be M-*complete* when every M-regular sequence in $\widetilde{X}$ has its limit in $\widetilde{X}$. Depending on M, $\widetilde{X}$ may or may not be M-complete. In many cases when X is the domain of a linear operator A it is possible to choose M so that $\widetilde{X}$ becomes an M-complete space to which A has a direct extension.

3 EXTENSIONS OF LINEAR OPERATORS IN HILBERT SPACE

Now let A be a densely defined linear operator in a Hilbert space H or, more precisely, assume that the domain of A is an inner product space whose Hilbert space completion contains the range of A. A vector x in H is called *a good element with respect to* A if x is in the domain of any polynomial in A and A^*. Thus the subspace G of good elements is invariant under A as well as under A^*.

The Hilbert space structure of H makes it possible to identify G with a subspace of G^f, or of H^f. It is convenient to have a notational distinction here, so I shall use G to denote the space of good elements with respect to an operator when these are identified with linear functionals on G or H. It is clear that G is total when G is dense in H, so that in this case we have the inclusions $G \subset H \subset \widetilde{G}(G) = \widetilde{H}(G)$, regarded as subspaces of G^f. Moreover, the transformation A has a direct extension $\widetilde{A}$ from $\widetilde{G}$ into $\widetilde{G}$. Whenever $\widetilde{x}$ is the limit† of a G-regular sequence (x_n) in G we simply define $\widetilde{A}\widetilde{x}$ to be the limit of (Ax_n). The equality $(Ax_n,y) = (x_n,A^*y)$ shows that (Ax_n) is G-regular, and also that the definition is consistent in the sense that $\widetilde{A}\widetilde{x}$ is well defined, independent of our choice of approximating sequence (x_n) to $\widetilde{x}$. The elements in $D(\widetilde{A}) = \widetilde{G}(G)$ may be regarded as *generalised elements with respect to* A.

In every special case that I have looked at, the extension $\widetilde{A}$ preserves the properties of the original operator A. The price we have to pay is, of course, that statements about $\widetilde{A}$ hold only in a »G-weak» sense. Furthermore, *the extended domain* $\widetilde{G}$ *has the important property of being G-*complete *when* A *is closed.* Among other things this implies that whenever a sequence $(\widetilde{x}_n)$ in $D(\widetilde{A}) = \widetilde{G}$ is G-regular (in the sense that $(\widetilde{x}_n - \widetilde{x}_m,y) \to 0$ as n and $m \to \infty$ for all y in G) then this sequence has a limit $\widetilde{x}$ in $D(\widetilde{A})$ and $\widetilde{A}\widetilde{x}_n$ converges to $\widetilde{A}\widetilde{x}$.

In the Hilbert space setting discussed here, the spaces of generalised elements with respect to linear operators may be obtained also by the distributional approach. It is possible to define a topology on G, or rather, in this context, on the representation G in G^f of this space, in such a way that $\widetilde{G}$ becomes precisely the space G' of continuous linear functionals on G. This topology is determined by the transformation A in such a way that G (or G) becomes a countably Hilbert space when A is closed. In fact, enumerating the 2^n terms in the expansion of $(A + A^*)^n$ for each natural number n, so that

† Throughout this talk all statements about limits, convergence, approximating sequences and the like should be interpreted in a G-regular (or G-weak) sense. Also, when $\widetilde{x}$ is in the subspace $\widetilde{G}$ of G^f and y is in G it seems natural to use $(\widetilde{x},y)$ rather than $< y,\widetilde{x} >$ to denote the value of $\widetilde{x}$ at y. Thus »(x_n) approximates to $\widetilde{x}$» means that $(x_n,y) \underset{n\to\infty}{\to} (\widetilde{x},y)$ for all y in G, etc.

$$(A + A^*)^n = \sum_{m=1}^{2^n} A_{nm} ,$$

we obtain a denumerable sequence of inner products on G by defining

$$(x,y)_p = \sum_{n=0}^{p} \sum_{m=1}^{2^n} (A_{nm}x, A_{nm}y) \qquad (p = 0,1,2,....).$$

The corresponding norms are increasing in p and pairwise compatible; they are used to define the desired metric topology on G just as in the case of the space S mentioned in the introduction. In the special case when A is self-adjoint the above countably Hilbert space G is

(i) *perfect* if and only if A has a pure point spectrum with no finite limit point,

(ii) *nuclear* if and only if G is perfect and the eigenvalues λ_ν (with multiplicity m_ν) for some integer p satisfy $\sum_{\lambda_\nu \neq 0} |\lambda_\nu|^{-p} m_\nu < \infty$.

4 EXTENSIONS OF SELF-ADJOINT OPERATORS

Assume now that T is a self-adjoint transformation acting in the Hilbert space H. The subspace of good elements with respect to T is then simply $G = D(T^\infty) = \bigcap_{n=1}^{\infty} D(T^n)$. The resolution of the identity E_λ associated with T maps G into G, and so it has, just like T, a direct extension $\widetilde{E}_\lambda : \widetilde{G} \to \widetilde{G}$ (here $\widetilde{G} = \widetilde{G}(G)$ is defined as above).It is simple to verify that $\widetilde{E}_\lambda$ is also a resolution of the identity, and that it has the same discontinuities and intervals of constancy as E_λ . In fact, the range of $\widetilde{E}_\lambda - \widetilde{E}_\mu$ (or $\widetilde{E}_{\lambda+0} - \widetilde{E}_\lambda$) is the same as that of $E_\lambda - E_\mu$ (or $E_{\lambda+0} - E_\lambda$). It follows that $\widetilde{T}$ and T have the same characteristic values and the same characteristic vectors. Moreover, μ is a point of continuity of $\widetilde{E}_\lambda$ if and only if $\widetilde{T} - \mu\widetilde{I}$ is injective and it is also a point of constancy if and only if $\widetilde{T} - \mu\widetilde{I}$ is surjective, here $\widetilde{I}$ denotes the identity map on $\widetilde{G}$.

Any operator function u(T) of T which maps G into G has an immediate extension $\widetilde{u(T)}$ from $\widetilde{G}$ into $\widetilde{G}$. This is the case if and only if the corresponding complex-valued function u on the real line satisfies

(i) u is locally in L^2_σ for each $\sigma(\lambda) = (E_\lambda x,x)$ with x in H,

(ii) the essential limit of $|\lambda|^{-p}u(\lambda)$ is zero as λ tends to infinity on the spectrum of T.

Other functions of T may be extended to well defined subsets of $\widetilde{G}$. The spectral theory for T carries over to $\widetilde{T}$, and operator Stieltjes integrals, constructed from operator Stieltjes sum in the usual way, may be used to define functions of $\widetilde{T}$ directly. For instance, when u satisfies (i) and (ii) above (so that $\widetilde{u(T)}$ is defined on all of $\widetilde{G}$) then

$\left(\int_{-n}^{n} u(\lambda)d\widetilde{E}_\lambda \widetilde{x}\right)_{n=1}^{\infty}$ is a G-regular sequence in G for each $\widetilde{x}$ in $\widetilde{G}$, with a limit $u(\widetilde{T})\widetilde{x} = \int_{-\infty}^{\infty} u(\lambda)d\widetilde{E}_\lambda\widetilde{x}$ (by notational definition) which is precisely $\widetilde{u(T)}\widetilde{x}$. In particular, the representations

$\widetilde{x} = \int_{-\infty}^{\infty} d\widetilde{E}_\lambda \widetilde{x}$ and $\widetilde{T}\widetilde{x} = \int_{-\infty}^{\infty} \lambda d\widetilde{E}_\lambda\widetilde{x}$ hold true for every $\widetilde{x}$ in $\widetilde{G}$. When the spectrum of T is discrete these representations take the form of (normally divergent) series expansions in the characteristic vectors of T.

The generalised elements with respect to T may also be thought of as the result of a repeated application of $\tilde{T}$ to a vector in H, plus a good element which has the character of a »constant» or »near constant» with respect to T. In applications where T is a differential operator this gives us a representation of $\tilde{G}$ as a class of derivatives, usually referred to as the »fundamental theorem of distribution theory». More precisely: given any generalised element $\tilde{x}$ (with respect to T) there exists a vector x in H, a vector y in G and a natural number n so that $\tilde{x} = \tilde{T}^n x + y$. When $\lambda = 0$ is not in the spectrum of T we may choose x so that $y = 0$. When $\lambda = 0$ is an isolated point of the spectrum we may choose x so that $Ty = 0$, and when $\lambda = 0$ is in the continuous spectrum so that, given any $\epsilon > 0$, $\|Ty\| < \epsilon$.

5 EXAMPLES

(1) The tempered distributions referred to in the introduction are the generalised elements with respect to the self-adjoint Hermite operator $T = -D^2 + t^2$, with domain and range in $L^2(-\infty, \infty)$. The result quoted at the end of section 3 implies that the corresponding countably Hilbert space S of good elements is nuclear. The standard spectral representation of generalised elements here takes the form of a series expansion in the Hermite functions φ_ν ($\nu = 0,1,2,\ldots$). For each $\tilde{x}$ in $\tilde{G}$ the sequence $\sum_{\nu=0}^{n} (\tilde{x}, \varphi_\nu)\varphi_\nu$ in G is G-regular and converges to $\tilde{x}$. The generalised elements are exactly those elements in G^f for which $\sum_{\nu=1}^{\infty} (\tilde{x}, \varphi_\nu)^2 \nu^{-p}$ converges for some natural number p. To be more precise, an element x′ in G^f is in $\tilde{G}$ if and only if $\sum_\nu < \varphi_\nu, x' >^2 \nu^{-p}$ converges for some p.

(2) Generalised functions on spaces of the form $K\{M_p\}$, with $M_p = q^p$ for some positive function $q : R^n \to R$ which tends to ∞ with $|x|$, are obtained if we choose a self-adjoint operator of the form $T = -\Delta + q$ with domain and range in R^n. Again the generalised elements may be represented by the standard series expansion $x = \sum_{\nu=0}^{\infty} (x, \varphi_\nu)\varphi_\nu$ in the characteristic vectors φ_ν of T, and $\tilde{G}$ consists precisely of those elements $\tilde{x}$ in G^f for which $\sum^{\infty} (\tilde{x}, \varphi_\nu)^2 \lambda_\nu^{-p} < \infty$ for some p.

(3) When $q : R^n \to R$ is defined by $q(x) = 1$ if $|x| < a$ and $q(x) = \infty$ if $|x| \geqslant a$, the space of good elements with respect to the transformation $T = -\Delta + q$ in $L^2(R^n)$ is the space $K(a)$ of infinitely differentiable functions on R^n which have their support contained in $|x| < a$. The union of the corresponding spaces of generalised functions for all natural numbers a is the standard space of distributions on C_0^∞.

(4) We may obtain summation methods for (arbitrarily rapidly) divergent series in any infinite orthonormal system $(\varphi_n)_{n=1}^{\infty}$ as follows: Choose a sequence (a_n) of real numbers tending to infinity, and define the »torsion» operator A by $D(A) = \left\{ \Sigma x_n \varphi_n;\ \Sigma |x_n a_n|^2 < \infty \right\}$ and then $A(\Sigma x_n \varphi_n) = \Sigma x_n a_n \varphi_{n+1}$. Then $\Sigma c_n \varphi_n$ represents a generalised element with respect to A if and only if there exists a natural number p such that

$$\sum_{n=1}^{\infty} \left| \frac{c_n}{a_n a_{n+1} \cdots a_{n+p}} \right|^2 < \infty.$$

(5) When T is the self-adjoint transformation generated by the differential operator iD in $L^2(-\infty,\infty)$ then every generalised element may be represented as

$$\tilde{x} = \frac{1}{\sqrt{2\pi}} \int_{-\infty}^{\infty} c(\lambda) e^{-i\lambda t} d\lambda,$$

with

$$c(\lambda) = \frac{1}{\sqrt{2\pi}} \frac{d}{d\lambda} \left(\tilde{x}, \frac{e^{-i\lambda t} - 1}{-it} \right),$$

i.e. $\left(\frac{1}{\sqrt{2\pi}} \int_{-n}^{n} c(\lambda) e^{-i\lambda t} d\lambda \right)_{n=1}^{\infty}$ is a *G*-regular sequence in G which converges to $\tilde{x}$. This generalises the Fourier-Plancherel transformation and its inverse; it identifies the generalised elements with respect to T with generalised elements with respect to multiplication by the independent variable. The sequence $\left(\int_{-n}^{n} c(\lambda) e^{-i\lambda t} d\lambda \right)$ converges to an element in $\tilde{G}$ if and only if there is an integer n for which $\int_{-\infty}^{\infty} \frac{|c(\lambda)|^2}{1+\lambda^{2n}} d\lambda$ converges.

(6) Let T be the differential operator iD restricted to those absolutely continuous functions f in $L^2(-a,a)$ which have their derivative in this space and satisfy $f(a) = f(-a)$. Then G is the space $\hat{K}(a)$ of infinitely differentiable functions of period 2a. The extension $\tilde{T}$ applied to an absolutely continuous function f on $(-a,a)$ does not, in general, give the result iDf. Instead we find that

$$\tilde{T}f = iDf - i\,[f(a) - f(-a)]\,\tilde{\delta}(t-a).$$

The extended operator »remembers» the boundary conditions, if we change these we get different extensions.

On functions holomorfic in tube domains $\mathbb{R}^n + iC$

E.M. de Jager

Abstract

In this paper we give a review of some well known properties of functions holomorfic in radial tube domains $\mathbb{R}^n + iC$, where C is a cone in $\mathbb{R}^n$. An application to solutions of a class of convolution equations is given. This application reveals the so called quasi analytic character of differences of these solutions: i.e. solutions which are equal in a domain G are also equal in a larger domain $B_C(G)$, the C-convex envelope of G. This property well known for solutions of the wave equation, appears to be of much more generality.

1. Introduction

The importance of the theory of functions of several complex variables for the theory of distributions has been established by investigations of a.o. L. Schwartz [6],[7] and J.L. Lions [5] concerning the support of Fourier transforms of functions holomorfic in C^n or $\mathbb{R}^n + iC$, where $\mathbb{R}^n$ is the space of the real variable $x = (x_1, x_2, \ldots, x_n)$, C^n the space of the complex variable $z = (z_1, z_2, \ldots, z_n)$ and C is an open cone in the space $\mathbb{R}^n$. Other investigations concerning functions of several complex variables and distributions deal with distributions as boundary values of holomorfic functions and we mention the researches by H. Bremermann [1], H.G. Tillmann [8]-[10], and R. Carmichael [4].

An other important relation is given by the now famous theorem of "the edge of the wedge" of Bogoliubov.

This theorem reads roughly as follows. If f is holomorfic in the domain $T_R^C = \{z = x+iy;\ |z| < R,\ y \in C\}$, with $C \cap -C \neq \emptyset$,

and if for all test functions ϕ, belonging to D(G),

$$\lim_{\substack{y\to 0\\ y\in C}} \int f(x+iy)\phi(x)dx \tag{1.1}$$

exists, independent of the path of y in C, then f is also holomorfic in the domain $T_R^C \cup \tilde{G}$, where $\tilde{G}$ is a complex neighbourhood of G.

The proof of this theorem is given by N.N. Bogoliubov, B.V. Medvedev, and M.K. Polivanov and also by F.E. Browder [3] and by H. Bremermann, R. Oehme and J.G. Taylor [2].
In this review we consider mainly functions holomorfic in tube domains $\mathbb{R}^n + iC$; we give a review of the above mentioned relations between holomorfic functions and distributions and also we give an application to solutions of a class of convolution equations. The application reveals the so called quasi analytic character of differences of these solutions, i.e. solutions which are equal in a domain G are in general also equal in a larger domain.
This property, well known for solutions of the wave equation, appears to be of much more generality in so far that solutions of the homogeneous wave equation are only a special example of quasi analytic distributions. This application is due to V.S. Vladimirov [12]; also the content of this review paper is based for a great deal on the work of this author [12].

2. Preliminary notions

As usual we denote by E the set of all functions, defined on $\mathbb{R}^n$ and having continuous derivatives of all orders.
The sets D and S are subsets of E with the respective properties that the functions out of D have compact supports and the functions out of S are strongly decreasing at infinity i.e.

$$x^pD^q\phi(x) = x_1^{p_1}x_2^{p_2}\dots x_n^{p_n} \frac{\partial^{|q|}}{\partial x_1^{q_1}\partial x_2^{q_2}\dots\partial x_n^{q_n}} \phi(x_1,x_2,\dots,x_n)$$

is uniformly bounded by a constant $C_{p,q}$, depending on the n-tuples p and q of non negative integers.
Using families of semi-norms the sets E, D and S are equiped with a topology (see e.g. [6],[11]), such that convergence in D implies convergence in S and convergence in S implies convergence in E ($D \subset S \subset E$).
The spaces E', D', S' are the spaces of all linear continuous forms (distributions) on respectively E, D and S.
The distributions in E' have compact support; the distributions in S' are generalized derivatives of continuous functions which do not increase faster than any power of $|x|$ for $|x|\to\infty$ (therefore they are also called

tempered distributions).
We have $E' \subset S' \subset D'$ and weak convergence in E' implies weak convergence in S' and weak convergence in S' implies weak convergence in D'.
We now define the classes H(a) and H(a;C) of holomorfic functions. The class H(a) is defined as the class of all functions, holomorfic in C^n, with the property

$$|f(z)| \leq C_{f,\varepsilon}(1+|z|)^m e^{(a+\varepsilon)|y|}, \tag{2.1}$$

where a is a fixed non negative, m an arbitrary non negative and ε an arbitrarely small positive number; the constant $C_{f,\varepsilon}$ depends on the function f and on the number ε, z is the n-tuple
$z = (z_1,z_2,\ldots,z_n) = (x_1+iy_1,x_2+iy_2,\ldots,x_n+iy_n) = x+iy, |z|^2 = \sum_{i=1}^{n} z_i\bar{z}_i$

and $|y|^2 = \sum_{i=1}^{n} y_i^2$.

The class H(a;C) consists of all functions f(z), which are holomorfic in $\mathbb{R}^n + iC$, where C is an open convex cone in $\mathbb{R}^n$ with apex in the origin and which have the following property:

$$|f(z)| \leq M(C')(1+|z|)^p(1+|y|^{-q})e^{a|y|} \tag{2.2}$$

for all y belonging to a cone C' with $\bar{C}' \subset C \cup \{0\}$; a is again a fixed non negative number and p and q are arbitrary non negative numbers, independent of the choice of C'; M is a constant which depends on f and on the choice of C' in C.
In the next section 3 we take for C either a cone C^+ lying in the upper half space $y_1 > 0$ or a cone C^- lying in the lower half space $y_1 < 0$.

A relation between these classes of holomorfic functions and certain classes of distributions can be brought about by means of the so called Fourier-Laplace transformation.
Let f(z) be a function holomorfic in the tube $T^B = \mathbb{R}^n + iB$, where B is a domain in $\mathbb{R}^n$. The spectral function of f(z) is defined as a distribution $g \in D'$ with the following properties:

a. $g(\xi)e^{-\xi\cdot y} \in S'$, $\forall y \in B$ and $\forall \xi \in \mathbb{R}^n$.

b. $f(z) = F[g(\xi)e^{-\xi.y}](x)$, where F denotes the Fourier transform. f(z) is called the Fourier-Laplace transform of the spectral function $g(\xi)$. The spectrum of f(z) is the support of the distribution $g(\xi)$.

The precise relation between functions belonging to the classes H(a) and H(a;C) and distributions belonging to E' respectively S' is given by the Paley-Wiener-Schwartz theorem valid for functions of the class H(a) and its extension valid for functions of the class H(a;C).
These theorems will be given in the next section.

3. The Paley-Wiener-Schwartz theorem.

Theorem 1. A function belonging to the class H(a) is the Fourier-Laplace transform of a spectral function $g(\xi)$, which belongs to $E'(\mathbb{R}^n)$ and which vabishes for $|\xi| > a$; also, conversely, the Fourier-Laplace transform of a distribution belonging to $E'(\mathbb{R}^n)$ and with support in a ball $|\xi| \leq a$ is an entire function belonging to the class H(a).

This theorem means that the relation (2.1) is equivalent with the relation

$$f(z) = F[g(\xi)e^{-y.\xi}](x) = \langle g(\xi), e^{ix.\xi - y\xi}\rangle$$
$$= \langle g(\xi), e^{iz\xi}\rangle, \text{ with } g = 0 \text{ for } |\xi|>a \quad (3.1)$$

For the proof of this well known theorem the reader is referred to lit. [6], Vol. 2, p. 128, where a short proof is given.

This theorem of Paley-Wiener-Schwartz has been extended for functions of the class H(a;C).
Before formulating this extension we mention the following lemma:

Lemma 1

A function of class H(a;C) possesses on S a weak limit as y approaches zero within C; i.e.

$$\lim_{\substack{y\to 0\\ y\in C}} \langle f(x+iy), \phi(x)\rangle = \langle f(x+i0), \phi(x)\rangle, \ \forall\phi \in S, \quad (3.2)$$

where the limit is independent of the path. (C in the upper half space $y_1>0$)
Hence the "boundary values" of functions belonging to H(a;C) are tempered distributions.

Proof: lit[12], p. 225.

The extension of the Paley-Wiener-Schwartz theorem may now be formulated as follows:

Theorem 2 A function f(z) of class H(a;C) is the Fourier-Laplace transform of a spectral function $g(\xi)$, which belongs to $S'(\mathbb{R}^n)$ and which vanishes for

$$\mu_C(\xi) \overset{\text{def}}{=\!=\!=} \sup_{\substack{y\in C\\ |y|=1}} (-\xi.y) > a \geq 0 \tag{3.3}$$

Also the inverse is true: the Fourier-Laplace transform of a distribution belonging to S' and vanishing for $\mu_C(\xi) > a \geq 0$, yields a function of the class H(a;C).

The spectral function is

$$g(\xi) = F^{-1}[f(x+i0)] \in S' \tag{3.4}$$

and the function f(z) may be represented as:

$$f(z) = \langle g(\xi), e^{i\xi z}\rangle = \langle g(\xi), e^{i\xi x-\xi y}\rangle \tag{3.5}$$

Remarks

1. The condition (3.3) is a natural restriction of the condition $|\xi| > a \geq 0$ of theorem 1, for

$$\sup_{\substack{y\in \mathbb{R}^n\\ |y|=1}} (-\xi.y) = |\xi| > a \geq 0$$

2. Let C' be a cone with closure in $C \cup \{0\}$ and let α be the radial distance between ∂C and $\partial C'$, then the scalar product $\xi.y = \sum_{i=1}^{n} \xi_i y_i$ satisfies for $\mu_C(\xi) \leq a$, ξ sufficiently large, and for $y \in C'$ the inequality:

$$\xi.y \geq |\xi|.|y| \sin(\partial C,\partial C') = |\xi|.|y| \sin\alpha > 0 \tag{3.6}$$

(see fig. 1). Hence $e^{i\xi x-\xi y} = e^{i\xi z}$ may be conceived as a test function of S for all $z = x+iy \in \mathbb{R}^n + iC$ and so the representation (3.5) is rather obvious.

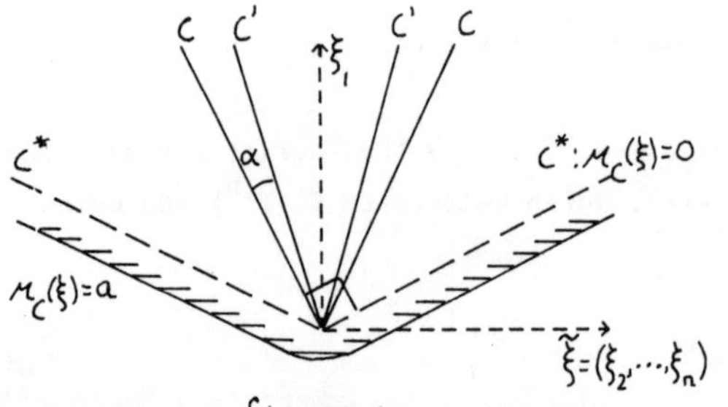

figure 1

<u>Proof of theorem 2:</u> lit [12], p.234.
See also lit. [7] and [5].

4. <u>The class $L_a(C)$ of quasi analytic functions.</u>

Let C be a cone which consists of two parts: one convex cone C^+ in the upper half space and one convex cone C^- in the lower half space with $C^+ = -C^-$. $C = C^+ \cup C^-$.

We introduce the class $L_a(C)$ of quasi analytic functions (a name which becomes clear later on) defined as:

$$L_a(C) = \{f \mid f \in S' \wedge (\text{Spectrum of } f) \subset F\} \tag{4.1}$$

with $F = F^+ \cup F^-$ and $F^{\pm} = \{\xi \mid \mu_{C^\pm}(\xi) \leq a\}$

Hence $L_a(C)$ consists of all distributions out of S', the inverse Fourier transforms of which have their supports in the hatched region of $\mathbb{R}^n(\xi)$. (See figure 2).

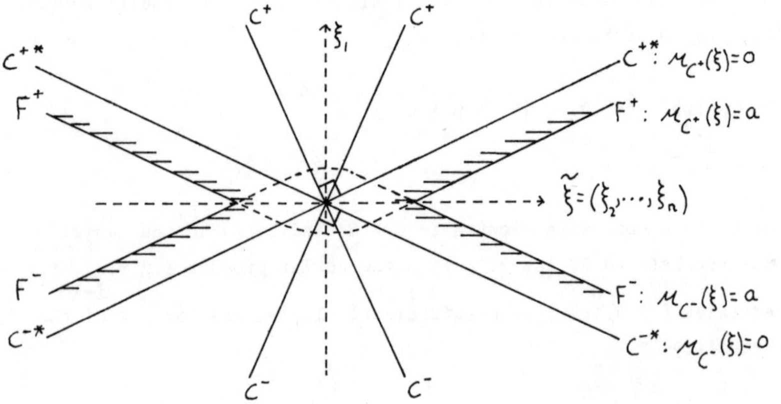

figure 2

The following theorem is now immediately obtained from the definition and the extended Paley-Wiener-Schwartz theorem of section 3.

Theorem 3. Any distribution of $L_a(C)$ allows a Hilbert-splitting, i.e. it can be represented as the difference of the boundary values of two functions, holomorfic in $\mathbb{R}^n + iC^+$ and $\mathbb{R}^n + iC^-$ and belonging to the respective classes $H(a;C^{\pm})$.

Hence $f \in L_a(C)$ implies

$$f(x) = f^+(x+i0) - f^-(x-i0) \qquad (4.2)$$

with $f^{\pm}(z) \in H(a;C^{\pm})$.

Also the inverse is true: (4.2) implies $f \in L_a(C)$.

Remarks

1. $f^{\pm}(z)$ is defined, apart from an entire function with spectrum in a ball with radius a.

2. When $f(x) \in L_a(C)$ and $f(x) = 0$ in an n-dimensionaal domain $G \subset \mathbb{R}^n$, then $f^+(z)$, holomorfic in $G+iC^+$, has an analytical continuation into $G+iC^-$; this analytical continuation is $f^-(z)$.

This statement is a consequence of the "edge of the wedge" theorem, mentioned in the introduction. For the proof of this theorem see lit. [3], [2] and [12 , p.241].

The most important property of the functions of class $L_a(C)$ is their quasi-analytic behaviour: whenever they are zero in a domain $G \subset \mathbb{R}^n$; then they are in general also zero in a larger convex domain $B_C(G) \subset \mathbb{R}^n$; hence $f^+(z)$ is not only holomorfic in a complex neighbourhood U of G, but also in a complex neighbourhood $\tilde{U}$ of $B_C(G)$.

This important property is derived from the following lemma:

Lemma 2 Let f(z) be holomorfic in $G + iC^+$ and $G + iC^-$ $(C^- = -C^+)$ and also holomorfic along a "time" like arc, lying in G and connecting the points $x^{(1)}$ and $x^{(2)}$; then f(z) is also holomorfic in $B_C(x^{(1)},x^{(2)})$.

A "time" like arc is an arc for which the tangent vector in any point P lies within the cone C^+ (or C^-) with P as apex. The domain $B_C(x^{(1)},x^{(2)})$ is the set of all points, which lie on "time" like arcs connecting $x^{(1)}$ and $x^{(2)}$. (See figure 3).

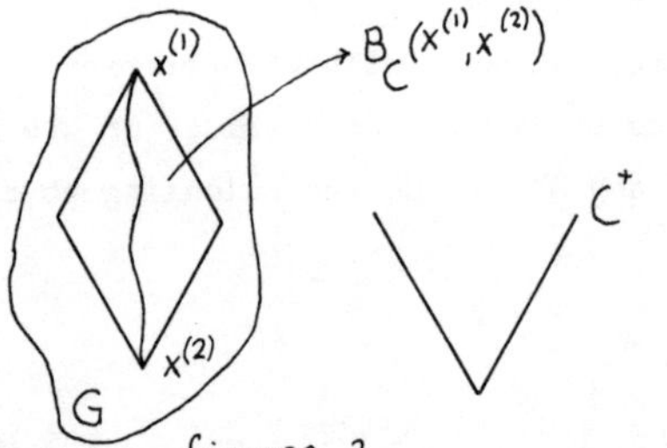

figure 3

<u>Proof</u> The proof is obtained by using a continuity argument and the so called theorem of the disk. (see lit. [12], p.261)

This lemma has the following immediate consequence:

<u>Theorem 4</u>. Let $f(x) \in L_a(C)$ and $f(x) = 0$ in a domain $G \in \mathbb{R}^n$, then $f(x) = 0$ in the C-convex envelope $B_C(G)$ of G, i.e. the set of all points lying on "time" like arcs connec ting any two points $x^{(1)}$ and $x^{(2)}$ in G. (See figure 4).

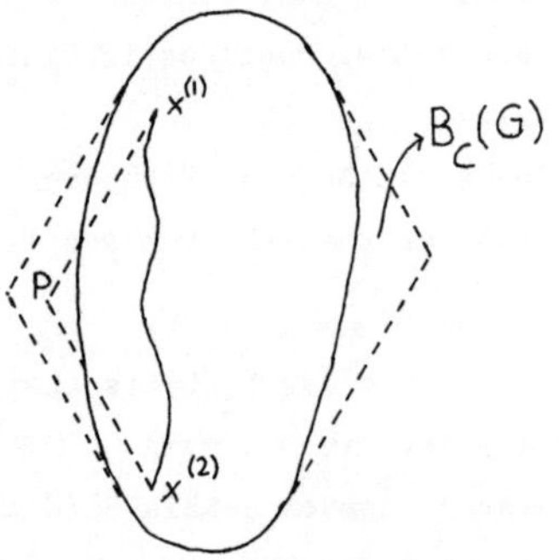

figure 4

<u>Proof</u>: According to theorem 3 $f(x)$ may be written as $f(x) = f^+(x+i0) - f^-(x-i0)$ with $f^{\pm}(z) \in H(a;C^{\pm})$.
Take $g(z) = f^+(z)$ for $y \in C^+$ and $g(z) = f^-(z)$ for $y \in C^-$. According to lemma 2, $f^+(z)$ possesses in P (see fig. 4) an analytic continuation into $\mathbb{R}^n + iC^-$, say $\tilde{f}^-(z)$.
We can also make this analytical continuation into $\mathbb{R}^n + iC^-$ via G, where $f^+(z)$ is holomorfic.

Hence $f^+(z+i0) = \tilde{f}^-(x-i0) = f^-(x-i0)$ in P, and thus
$f(x) = f^+(x+i0) - f^-(x-i0) = 0$ in P.

5. <u>Application to convolution equations</u>.

An an application we mention the following theorem:

<u>Theorem 5</u>. Let $u \in S'$ be a solution of the convolution equation

$$f_0 * u = f_1 \qquad (5.1)$$

with $f_1 \in L_a(C)$ and $f_0 \in 0_{conv}$ (i.e. f_0 belongs to the set of all convolutors in S').

If $F^{-1}[f_0](\xi) \neq 0$ for $\min[\mu_{C^+}(\xi), \mu_{C^-}(\xi)] > a$ (5.2)

then $u \in L_a(C)$.

The proof follows immediately from the equation:

$$F^{-1}[f_0].F^{-1}[u] = F^{-1}[f_1]$$

A consequence of theorem 5 is: when two distributions u_1 and u_2, belonging to S', satisfy the convolution equation

$$f_0 * u = g \tag{5.3}$$

with f_0, submitted to the condition (5.2), and with g a given distribution of S', then their difference $u_1 - u_2$ belongs to $L_a(C)$.

When two solutions u_1 and u_2 of (5.3) concide on a domain $G \subset \mathbb{R}^n$, then they coincide also on $B_C(G)$ (see theorem 4).

Examples

1. Differential equations.

$P(iD)u(x) = g(x)$, with P a polynomial in $iD = i(\frac{\partial}{\partial x_1}, \frac{\partial}{\partial x_2}, \ldots, \frac{\partial}{\partial x_n})$,

$f_0 = P(iD)\delta(x)$ and $F^{-1}[f_0](\xi) = \frac{1}{(2\pi)^n} P(-\xi)$; (5.2) means $P(-\xi) \neq 0$ for $\xi \in \{\xi \mid (\mu_{C^+}(\xi) > a) \wedge (\mu_{C^-}(\xi) > a)\}$.

Two solutions coinciding on G, coincide also on $B_C(G)$.

This property is well known for solutions of the wave equation, which is a special example with $P(\xi) = \xi_1^2 - \xi_2^2 - \ldots - \xi_n^2$, C^+ the forward and C^- the backward light cone and a = 0.

2. Difference equations.

$\sum_1^N c_k u(x - x^{(k)}) = g(x)$, with c_k given constants and $x^{(k)}$ given points in $\mathbb{R}^n$. $f_0 = \sum_1^N c_k \delta(x - x^{(k)})$ and $F^{-1}[f_0](\xi) = \frac{1}{(2\pi)^n} \sum_1^N c_k e^{-ix^{(k)}.\xi}$

3. Integral equations.

$$\int_{-\infty}^{+\infty} K(x-x')u(x')dx' = g(x).$$

$f_0(x) = K(x)$ with $K(x)$ e.g. a continuous function, decreasing at infinity stronger than any negative power of $|x|$.

$$F^{-1}[f_0](\xi) = \frac{1}{(2\pi)^n} \int_{-\infty}^{+\infty} K(x)e^{-ix.\xi}d\xi.$$

Literature

1. H. Bremermann, Distributions, Complex Variables, and Fourier Transforms; Addison-Wesley, Reading, 1965.

2. H. Bremermann, R. Oehme, J.G. Taylor, A proof of dispersion relations in quantized field theories; Phys. Rev. 109 (1958), p. 2178.

3. F.E. Browder, On the "edge of the wedge" theorem; Canad. J. of Mathematics, 15 (1963), p. 125.

4. R. Carmichael, Generalized Cauchy and Poisson integrals and distributional boundary values; SIAM J. of Math. Anal., 4, No. 1 (1973), p. 198.

5. J.L. Lions, Supports dans la transformation de Laplace; J. Analyse Math., 2, (1952-1953), p. 369.

6. L. Schwartz, Théorie des distributions, Tome 1,2; Hermann, Paris, (1957) resp. (1959).

7. L. Schwartz, Transformation de Laplace des distributions; Medd. Lund Univ., mat. semin., (1952), pp. 196-206.

8. H.G. Tillmann, Randverteilungen analytischer Funktionen und Distributionen; Math. Zeitschrift, 59, (1953),p. 61.

9. H.G. Tillmann, Distributionen als Randverteilungen analytischer Funktionen II; Math. Zeitschrift, 76, (1961), p. 5.

10. H.G. Tillmann, Darstellung der Schwartzschen Distributionen durch analytische Funktionen; Math. Zeitschrift, 77, (1961), p. 106.

11. F. Trèves, Topological Vector Spaces, Distributions and Kernels; Academic Press, New-York, (1967).

12. V.S. Vladimirov, Les fonctions de plusieurs variables complexes et leur application à la théorie quantique des champs; Dunod, Paris, (1967).

ON THE ASYMPTOTIC THEORY OF NON-LINEAR OSCILLATIONS.

WIKTOR ECKHAUS
Mathematisch Instituut, Rijksuniversiteit te Utrecht
Utrecht, The Netherlands

1. INTRODUCTION.

This paper reports on a recently developed new approach to the asymptotic theory of non-linear oscillations and wave propagation. Only the main results will be discussed here, full details of the theory will be presented in a forthcoming publication (Eckhaus (1974)).

The starting point of our analysis is the well known asymptotic method of Krilov-Bogolioubov-Mitropolski, see for example Bogolioubov and Mitropolski (1961)). To define the ideas let us briefly outline some of the fundamental results.

Suppose $\underline{Y}(t,\varepsilon)$ is an n-dimensional vector function of time-like variable t and a small parameter ε. $\underline{Y}(t,\varepsilon)$ is defined as solution of the initial value problem

$$\frac{d\underline{Y}}{dt} = \varepsilon \underline{F}(\underline{Y},t,\varepsilon),\ \underline{Y}(0,\varepsilon) = \underline{Y}_0$$

where $\underline{F}$: $\mathbb{R}^n \times \mathbb{R} \times \mathbb{R} \to \mathbb{R}^n$.
According to the asymptotic method $\underline{Y}(t,\varepsilon)$ can be approximated by a function $\underline{\eta}(\varepsilon t)$, defined as solution of

$$\frac{d\underline{\eta}}{dt} = \varepsilon \underline{F}_0(\underline{\eta})\ ;\ \eta(0) = \underline{Y}_0$$

where

$$\underline{F}_0(\underline{\eta}) = \lim_{T\to\infty} \frac{1}{T} \int_0^T \underline{F}(\underline{\eta},t,0)dt$$

The approximation is valid in the sense that $|\underline{Y}(t,\varepsilon) - \underline{\eta}(\varepsilon t)|$ tends to zero as ε tends to zero. The validity is assured in an interval

$$0 \leqslant t \leqslant \frac{L}{\varepsilon}$$

where L is an arbitrary constant (independent of ε). In the special case of periodic systems, that is $\underline{F}(\underline{Y},t+T,\varepsilon) = \underline{F}(\underline{Y},t,\varepsilon)$ for some constant T, a better estimate can be obtained. In that case one can show that $|\underline{Y}(t,\varepsilon) - \underline{\eta}(\varepsilon t)| \leqslant \varepsilon c(L)$, where c(L) is a constant which only depends on L. Furthermore, higher approximations of $\underline{Y}(t,\varepsilon)$ can also be defined.
There are several reasons which have led the present author to recinsider this well-known and well-developed theory :
The proofs given in any standard text, are difficult.

Furthermore, in any standard presentation, the theory is not deductive. By this we mean that without any convincing motivation one asserts that $\underline{\eta}$ is an approximation of $\underline{Y}$, and one proceeds to proof the assertion by classical procedure of estimation of integrals. There is little in the development of the theory (except for the results) that justifies the name "asymptotic method".

Finally, one wonders whether it could be possible to extend the validity of the results to larger time-intervals. Such extension has sofar only been established for the case of periodic solutions or solutions starting sufficiently near stable periodic solutions.
In the study described in this paper a deductive asymptotic theory is developed, which uses from the outset concepts and methods of asymptotic analysis. The necessary preliminaries are given in section 2. Section 3 introduces the fundamental tool of our method of analysis: in a suitable (asymptotic) sense a local average value of the function $\underline{Y}$ is defined. With the aid of this concept a deductive procedure establishes the fundamental theory of Krilov -Bogolioubov-Mitropolski under the most general conditions (section 4).
In the final section we report on some new results :
For a class of problems validity of the asymptotic approximation on $0 \leqslant t < \infty$ has been established. Furthermore, application to partial differential equations shows that the concept of local average values can be very useful tool in studying wave-propagation phenomena.
A remark on the formulation used in this paper should be made. Throughout our analysis we study vector equations

$$\frac{d\underline{Y}}{dt} = \varepsilon\,\underline{F}(\underline{Y},t,\varepsilon)$$

which we have termed "standard systems for slowly modulated processes". A large class of problems can be transformed into this form. Well-known example is given by systems of perturbed linear oscilators:

$$\frac{d^2x_i}{dt^2} + w_i^2x_i = \varepsilon g_i(x_1,\ldots x_m, \frac{dx_1}{dt},\ldots \frac{dx_m}{dt},t);\; i = 1,\ldots,m$$

If now $\underline{X}$ is the vector with components $x_1,\ldots x_n$, then the correspondence $\underline{X} \to \underline{Y}$ is achieved by the Van der Pol transformation. However, also non-linear perturbed systems, of the general form

$$\frac{d\underline{z}}{dt} = \underline{H}(\underline{z},t) + \varepsilon\underline{G}(z,t)$$

can, under certain conditions, be transformed into a slowly modulated standard system. Raughly speaking this is possible when the "unperturbed system"

$$\frac{d\underline{z}^0}{dt} = \underline{H}(\underline{z}^0,t)$$

possesses a "general solution". The transformation is then in essence achieved by the method of variation of constants. The conditions that arise in the course of the calculations can be found in Volosov (1962).

2. ELEMENTARY PROPERTIES OF SLOWLY MODULATED STANDARD SYSTEMS.

We study vector functions $\underline{Y}(t,\varepsilon)$ which are defined as solutions of the initial value problem

$$\frac{dY}{dt} = \varepsilon \underline{F}(\underline{Y},t,\varepsilon) \ ; \ \underline{Y}(0,\varepsilon) = \underline{Y}_0$$

where $\underline{F} : \mathbb{R}^n \times \mathbb{R} \times \mathbb{R} \to \mathbb{R}^n$ is a vector function with components $f_i(y_1,\dots,y_n, t,\varepsilon)$, defined in some connected subset $G \subset \mathbb{R}^{n+2}$.

The above differential equation, which represents a system of n first order differential equations for the components $y_1,\dots,y_n$ of $\underline{Y}$, will be called a <u>standard system for slowly modulated processes</u>.

We shall suppose throughout our analysis that the function $\underline{F}$ satisfies the following conditions :

i. $\underline{F}$ is a continuous and uniformly bounded function in G, where

$$G = \{\underline{Y} | \underline{Y} \in \bar{D}\} \times \{t | 0 \leqslant t < \infty\} \times \{\varepsilon | 0 \leqslant \varepsilon \leqslant \varepsilon_0\}$$

D is some open bounded subset of $\mathbb{R}^n$.

$$\lim_{\varepsilon \to 0} \{\underline{F}(\underline{Y},t,\varepsilon) - \underline{F}(\underline{Y},t,0)\} = 0 \text{ uniformly in}$$

$$\{\underline{Y} | \underline{Y} \in \bar{D}\} \times \{t | 0 \leqslant t < \infty\}$$

ii. $\underline{F}$ is Lipschitz-continuous with respect to $\underline{Y}$ in G, that is: there exists a constant λ such that for any pair $(\underline{Y}_1,\underline{Y}_2) \in D$ we have in G

$$|\underline{F}(\underline{Y}_1,t,\varepsilon) - \underline{F}(\underline{Y}_2,t,\varepsilon)| \leqslant \lambda |\underline{Y}_1 - \underline{Y}_2|$$

<u>Remarks.</u> The above conditions are for the most part the classical conditions needed to assure existence and uniqueness of the solution $\underline{Y}(t,\varepsilon)$. In addition, uniform behaviour as $\varepsilon \downarrow 0$ appears as a necessary condition for the existence of asymptotic approximations valid for $\varepsilon \downarrow 0$. Finally, uniform boundness on the whole time-axis $0 \leqslant t < \infty$ has been imposed to assure existence of solutions on a sufficiently large time interval. This becomes apparent from the following result.

<u>Lemma 1.</u> If $\underline{F}$ satisfies conditions i and ii and $\underline{Y}_0 \in D$, then there exists a unique solution $\underline{Y}(t,\varepsilon)$ of

$$\frac{d\underline{Y}}{dt} = \varepsilon \underline{F}(\underline{Y},t,\varepsilon) \ ; \ \underline{Y}(0,\varepsilon) = \underline{Y}_0, \text{ in } 0 \leqslant t \leqslant T \text{ with}$$

$$T = \frac{d}{\varepsilon M}$$

where d is the distance of $\underline{Y}_0$ to the boundary of D, and M is defined by

$$M = \sup_G |\underline{F}|$$

The proof of Lemma 1 is obtained by an almost trivial

modification of the classical existence and uniqueness theorem, as given for example in Roseau (1966).

By virtue of Lemma 1 it is meaningful to study $\underline{Y}(t,\varepsilon)$ on the natural time scale ε^{-1}. Introduce therefore : $\tau = \varepsilon t$

writing $\underline{Y}^*(\tau,\varepsilon) = \underline{Y}(\frac{\tau}{\varepsilon},\varepsilon)$ we have the initial value problem

$$\frac{d\underline{Y}^*}{d\tau} = \underline{F}(Y^*, \frac{\tau}{\varepsilon},\varepsilon) \ ; \ \underline{Y}^*(0,\varepsilon) = \underline{Y}_0 .$$

Unique solution exists in some interval $0 \leqslant \tau \leqslant T^*$ where $T^* > 0$ is a number independent of ε.

Continuation of the solution can be obtained by the following classical corollary of the existence and uniqueness theorem :

Lemma 2. Let I be a closed interval such that for $\tau \in I$ unique solution $\underline{Y}^*(\tau,\varepsilon)$ exists and $\underline{Y}^*(\tau,\varepsilon) \in K$ where K is a compact subset of D. Then a unique continuation of $\underline{Y}^*(\tau,\varepsilon)$ exists in some open interval containing I. Furthermore, the solution $\underline{Y}^*(\tau,\varepsilon)$ may be continued to all values of $\underline{\tau}$, for which the continuation remains in a compact subset of D.

In our analysis the continuation of the solution $\underline{Y}^*(\tau,\varepsilon)$ will be obtained from consideration of approximate solutions. For the study of approximate solutions, valid as $\varepsilon \downarrow 0$, a general approximation theorem can be formulated. As a preliminary we remind the reader that the initial value problem is equivalent with the integral equation

$$\underline{Y}^*(\tau,\varepsilon) = \underline{Y}_0 + \int_0^\tau \underline{F}[\,\underline{Y}^*(\tau',\varepsilon), \frac{\tau'}{\varepsilon}, \varepsilon]d\tau'$$

As a second preliminary the defination of special subsets of D will be given, because such subset will frequently appear in the subsequent analysis.

Definition

$D_0 \subset D$ is an interior subset if the distance between the boundary of D_0 and the boundary of D is bounded from below by a positive constant, independent of ε, for all $0 \leqslant \varepsilon \leqslant \varepsilon_0$.

We now have :

Theorem I.

Consider two functions $\underline{Y}^{(1)}$ and $\underline{Y}^{(2)}$,

$$\underline{Y}^{(1)}(\tau,\varepsilon) = \underline{Y}_0^{(1)} + \int_0^\tau \underline{F}_1[\,\underline{Y}^{(1)}(\tau',\varepsilon),\frac{\tau'}{\varepsilon}, \varepsilon]\, d\tau'$$

$$\underline{Y}^{(2)}(\tau,\varepsilon) = \underline{Y}_0^{(2)} + \int_0^\tau \underline{F}_2[\,\underline{Y}^{(2)}(\tau',\varepsilon),\frac{\tau'}{\varepsilon}, \varepsilon]\, d\tau'$$

Suppose :

i. $\underline{Y}_0^{(1)} \in D_0$, $\underline{Y}_0^{(1)} \in D_0$, $|\underline{Y}_0^{(1)} - \underline{Y}_0^{(2)}| \leqslant \delta_0(\varepsilon)$, $\delta_0(\varepsilon) = o(1)$

ii. For all $\underline{Y} \in \overline{D}$ and $0 \leqslant \tau \leqslant A$

$$|F_1(\underline{Y}, \frac{\tau}{\varepsilon}, \varepsilon) - F_2(\underline{Y}, \frac{\tau}{\varepsilon},\varepsilon)| \leqslant \delta_f(\varepsilon),\ \delta_f(\varepsilon) = o(1)$$

iii. Solution $\underline{Y}^{(2)}(\tau,\varepsilon)$ exists for $0 \leqslant \tau \leqslant A$ and $\underline{Y}^{(2)} \in D_0$.
Then : Solution $\underline{Y}_0^{(1)}(\tau,\varepsilon)$ exists for $0 \leqslant \tau \leqslant A$ and in that interval

$$|\underline{Y}^{(1)}(\tau,\varepsilon) - \underline{Y}^{(2)}(\tau,\varepsilon)| \leqslant \delta_0(\varepsilon)\, e^{\lambda\tau} + \delta_f(\varepsilon)\, \frac{1}{\lambda}\, (e^{\lambda\tau} - 1)$$

For the proof of theorem I we shall use Gronwall's lemma in a form given for example in Coddington and Levinson (1955) :

Lemma 3. Let $\lambda(\tau) \geqslant 0$ be an integrable function, while $u(\tau)$ and $\phi(\tau)$ are absolutely continuous functions for $\tau_0 \leqslant \tau \leqslant \tau_1$.

If $u(\tau) \leqslant \phi(\tau) + \int_{\tau_0}^{\tau} \lambda(\tau')\, d\tau'$

then

$$u(\tau) \leqslant \phi(\tau_0) \exp \int_{\tau_0}^{\tau} \lambda(\tau')\, d\tau' +$$

$$+ \int_{\tau_0}^{\tau} \frac{d\phi(\tau'')}{d\tau''} \left[\exp \int_{\tau''}^{\tau} \lambda(\tau')d\tau'\right] d\tau''$$

Proof of Theorem I.

$$|\underline{Y}^{(1)}(\tau,\varepsilon) - \underline{Y}^{(2)}(\tau,\varepsilon)| \leqslant$$

$$\delta_0(\varepsilon) + \int_0^{\tau} |F_1[\underline{Y}^{(1)}(\tau',\varepsilon), \frac{\tau'}{\varepsilon},\varepsilon] - F_2[\underline{Y}^{(2)}(\tau',\varepsilon),\frac{\tau'}{\varepsilon},\varepsilon]\, |d\tau' \leqslant$$

$$\delta_0(\varepsilon) + \int_0^{\tau} |F_1[\underline{Y}^{(1)}(\tau',\varepsilon), \frac{\tau'}{\varepsilon}, \varepsilon] - F_1[\underline{Y}^{(2)}(\tau',\varepsilon)\frac{\tau'}{\varepsilon},\varepsilon]\,|d\tau' +$$

$$\int_0^{\tau} |F_1[\underline{Y}^{(2)}(\tau',\varepsilon),\frac{\tau'}{\varepsilon},\varepsilon] - F_2[\underline{Y}^{(2)}(\tau',\varepsilon),\frac{\tau'}{\varepsilon},\varepsilon]\,|d\tau'$$

Since $\underline{Y}^{(1)}(\tau,\varepsilon)$ is a continuous function, $\underline{Y}^{(1)}$ will certainly remain in D_0 for some $0 \leqslant \tau \leqslant \tau_1$. We can therefore use Lipschitz-continuity, and furthermore property ii. We obtain in $0 \leqslant \tau \leqslant \tau_1$

$$|\underline{Y}^{(1)}(\tau,\varepsilon)-\underline{Y}^{(2)}(\tau,\varepsilon)| \leqslant$$

$$\leqslant \delta_0(\varepsilon) + \delta_f(\varepsilon)\, \tau + \lambda \int_0^{\tau} |\underline{Y}^{(1)}(\tau',\varepsilon) - \underline{Y}^{(2)}(\tau',\varepsilon)|d\tau'$$

Using Gronwall's lemma we find in $0 \leqslant \tau \leqslant \tau_1$

$$|\underline{Y}^{(1)}(\tau,\varepsilon) - \underline{Y}^{(2)}(\tau,\varepsilon)| \leqslant \delta_0(\varepsilon)\, e^{\lambda\tau} + \delta_f(\varepsilon)\, \frac{1}{\lambda}\, (e^{\lambda\tau} - 1)$$

From Lemma 2 we obtain a contnuation of the solution $\underline{Y}^{(1)}(\tau,\varepsilon)$, which remains in D_0 for sufficiently small ε. For every continuation the above estimate of $|\underline{Y}^{(1)} - \underline{Y}^{(2)}|$ remains valid, as long as $\underline{Y}^{(2)}$ remains in D_0. Hence the continuation and the estimate are valid in $0 \leqslant \tau \leqslant A$.

3. LOCAL AVERAGE VALUES.

When studying the initial value problem on the natural time scale

$$\frac{d\underline{Y}^*}{d\tau} = \underline{F}(\underline{Y}^*, \frac{\tau}{\varepsilon}, \varepsilon) \; ; \; \underline{Y}^*(0,\varepsilon) = \underline{Y}_0$$

one is often confronted with the case in which $\lim_{\varepsilon \to 0} \underline{F}(\underline{Y}^*, \frac{\tau}{\varepsilon}, \varepsilon)$ does not exist. Such is for example the case when $\underline{F}(\underline{Y},t,\varepsilon)$ is a periodic function of t, with period T independent of ε. Nevertheless, as is well-known from the Krilov-Bogolioubov theory, an asymptotic approximation $\underline{\eta}(\tau)$ of $\underline{Y}(\tau,\varepsilon)$ may exist. In our analysis the essential tool for the study of such problems is the concept of local average values that will be introduced now.

Definition 2.

Consider a function $(t,\varepsilon) \to \underline{\Phi}(t,\varepsilon)$ and a transformation $\tau = \delta_s(\varepsilon)t$, $\underline{\Phi}(\frac{\tau}{\delta_s},\varepsilon) = \underline{\Phi}^*(\tau,\varepsilon)$. A local average value $\underline{\overline{\Phi}}(\tau,\varepsilon)$ of $\underline{\Phi}(t,\varepsilon)$ on the time scale δ_s^{-1} is given by

$$\underline{\overline{\Phi}}(\tau,\varepsilon) = \frac{1}{\delta(\varepsilon)} \int_0^{\delta(\varepsilon)} \underline{\Phi}^* (\tau+\tau',\varepsilon)\, d\tau'$$

where δ is some order function with $\delta(\varepsilon) = o(1)$.

Remarks.

In the definition above the function $\underline{\Phi}(t,\varepsilon)$ is in fact averaged (in the usual sense of the word) over a "small" distance on the δ_s^{-1} time scale. The "smallness" of the distance over which the averaging is performed is in asymptotic sense, and is measured by the order function $\delta(\varepsilon)$. It is obvious that a "small" distance on the δ_s^{-1} time scale may be a "large" distance in the original t time variable. The average $\underline{\overline{\Phi}}(\tau,\varepsilon)$ depends on the choice of $\delta(\varepsilon)$, which leaves us with a degree of liberty to be exploited later on in the analysis. Naturally, $\underline{\overline{\Phi}}(\tau,\varepsilon)$ also depends on the time scale δ_s^{-1}, on which $\underline{\Phi}(t,\varepsilon)$ is being investigated.
The asymetry in the definition of $\underline{\overline{\Phi}}(\tau,\varepsilon)$ ("forward" integration, $t' \geqslant \tau$) is chosen, because otherwise $\underline{\overline{\Phi}}(0,\varepsilon)$ could not be defined. Finally we remark that for the purpose of calculation it is often advantageous to introduce an obvious change of the integration variable $\tau' = \delta\overline{\tau}$, which yields

$$\underline{\overline{\Phi}}(\tau,\varepsilon) = \int_0^1 \underline{\Phi}^*(\tau + \delta\overline{\tau},\varepsilon)\, d\overline{\tau}$$

The usefulness of the local average values immediately appears from the following fundamental result on the natural time scale $\tau = \varepsilon t$:

Lemma 4. Let $\underline{Y}^*(\tau,\varepsilon)$ be the solution in $0 \leqslant \tau \leqslant A$ of

$$\frac{d\underline{Y}^*}{d\tau} = \underline{F}(\underline{Y}^*, \frac{\tau}{\varepsilon},\varepsilon) \; ; \; \underline{Y}^*(0,\varepsilon) = \underline{Y}_0$$

then

$$\underline{Y}^*(\tau,\varepsilon) = \overline{\underline{Y}}(\tau,\varepsilon) + O(\delta).$$

For the proof of Lemma 4 we need, as a preliminary, a result which although elementary will be stated in a seperate Lemma because it will frequently be used in the sequel.

Lemma 5. Let $\underline{Y}^*(\tau,\varepsilon)$ be defined as in Lemma 4. Then

$$|\underline{Y}^*(\tau + \delta\overline{\tau},\varepsilon) - \underline{Y}^*(\tau,\varepsilon)| \leqslant M\delta\overline{\tau}$$

where $M = \sup_G |\underline{F}|$.

Proof of Lemma 5.

From the differential equation it follows that

$$|\underline{Y}^*(\tau+\delta\overline{\tau},\varepsilon) - \underline{Y}^*(\tau,\varepsilon)| \leqslant \int_{\tau}^{\tau+\delta\overline{\tau}} |\underline{F}(\underline{Y}^*(\tau',\varepsilon), \frac{\tau'}{\varepsilon},\varepsilon)|d\tau' \leqslant M\delta\overline{\tau}$$

Proof of Lemma 4.

From the definition of the average values we have

$$\overline{\underline{Y}}(\tau,\varepsilon) = \int_0^1 \underline{Y}^*(\tau+\delta\overline{\tau},\varepsilon)d\overline{\tau} = \underline{Y}^*(\tau,\varepsilon) + \int_0^1 \{\underline{Y}^*(\tau+\delta\overline{\tau},\varepsilon)-\underline{Y}^*(\tau,\varepsilon)\}\, d\overline{\tau}$$

Using now Lemma 5 we obtain

$$|\overline{\underline{Y}}(\tau,\varepsilon) - \underline{Y}^*(\tau,\varepsilon)| \leqslant \tfrac{1}{2}M\delta,$$

which proofs lemma 4.

Finally we investigate the relation between the local averages in the sense of definition 2, and the classical averages of Krilov and Bogolioubov defined by :

Definition 3.

$\underline{F}(\underline{Y},t,\varepsilon)$ is a K.-B. function if

$$\underline{F}_0(\underline{Y},\varepsilon) = \lim_{T\to\infty} \frac{1}{T} \int_0^T \underline{F}(\underline{Y},t,\varepsilon)dt \text{ exists.}$$

Now, in the sense of definition 2, a local average value of $\underline{F}(\underline{Y}^*,\frac{\tau}{\delta_s},\varepsilon)$ is given by $\overline{\underline{F}}(\underline{Y}^*,\frac{\tau}{\varepsilon},\varepsilon) = \frac{1}{\delta(\varepsilon)} \int_0^{\delta(\varepsilon)} \underline{F}(Y^*,\frac{\tau}{\delta_s}+\frac{\tau'}{\delta_s},\varepsilon)d\tau'$

We shall now prove the following correspondence :

Lemma 6. If $\underline{F}$ is a K.-B. function then there exists an order function $\delta_1(\varepsilon) = o(1)$ such that for any choice of $\delta(\varepsilon)$ in the averaging process, satisfying

$$\frac{\delta_1}{\delta} = o(1), \; \frac{\delta_s}{\delta} = o(1)$$

we have, uniformly on any bounded closed interval of the τ variable

$$\overline{\underline{F}}(\underline{Y}^*, \frac{\tau}{\varepsilon},\varepsilon) = \underline{F}_0(\underline{Y}^*,\varepsilon) + o(1)$$

Proof.

In definition 3 and Lemma 6 the variable $\underline{Y}^*$ only appears as a parameter. We therefore write, to simplify the notation

$$\underline{F}(Y,t,\varepsilon) = \underline{\Phi}(t,\varepsilon) \ ; \ \underline{F}_0(\underline{Y},\varepsilon) = \underline{\Phi}_0 .$$

If $\underline{\Phi}(t,\varepsilon)$ is a finite sum of functions periodic with respect to t (and this is the case which most often occurs in applications) then the proof of Lemma 6 can be obtained from straightforward computation of the average $\overline{\underline{\Phi}}(\tau,\varepsilon)$.

One then finds (see Eckhaus (1974)) :

$$\overline{\underline{\Phi}}(\tau,\varepsilon) - \underline{\Phi}_0 = O\left(\frac{\delta_s}{\delta}\right)$$

In the most general case, that is without supposing periodicity of $\underline{\Phi}(t,\varepsilon)$, the proof is somewhat more involved.

We are given that

$$\lim_{T\to\infty} \frac{1}{T} \int_0^T \{\underline{\Phi}(t,\varepsilon) - \underline{\Phi}_0\}dt = 0$$

Hence there exists a positive, continuous, monotonic function $\phi(z)$ with the property

$$\lim_{z\to 0} \phi(z) = 0$$

such that, for say $T > T_0$:

$$\left|\frac{1}{T} \int_0^T \{\underline{\Phi}(t,\varepsilon) - \underline{\Phi}_0\}dt\right| \leqslant \phi\left(\frac{1}{T}\right)$$

On the other hand, by definition 2 :

$$\overline{\underline{\Phi}}(\tau,\varepsilon) = \int_0^1 \underline{\Phi}\left(\frac{\tau}{\delta_s} + \frac{\delta}{\delta_s}\tau',\varepsilon\right)d\tau' = \frac{\delta_s}{\delta} \int_{\frac{\tau}{\delta_s}}^{\frac{1}{\delta_s}(\tau+\delta)} \underline{\Phi}(t,\varepsilon)dt$$

We now write

$$\overline{\underline{\Phi}}(\tau,\varepsilon) - \underline{\Phi}_0 = \frac{\delta_s}{\delta} \int_0^{\frac{1}{\delta_s}(\tau+\delta)} \{\underline{\Phi}(t,\varepsilon) - \underline{\Phi}_0\}dt -$$

$$- \frac{\delta_s}{\delta} \int_0^{\frac{1}{\delta_s}\tau} \{\underline{\Phi}(\tau,\varepsilon) - \underline{\Phi}_0\}dt$$

Consequently

$$|\overline{\underline{\Phi}}(\tau,\varepsilon) - \underline{\Phi}_0| \leqslant J_1 + J_2$$

$$J_1 = \left|\frac{\delta_s}{\delta} \int_0^{\frac{1}{\delta_s}(\tau+\delta)} \{\underline{\Phi}(t,\varepsilon) - \underline{\Phi}_0\}dt\right|$$

$$J_2 = \left|\frac{\delta_s}{\delta} \int_0^{\frac{1}{\delta_s}\tau} \{\underline{\Phi}(t,\varepsilon) - \underline{\Phi}_0\}dt\right|$$

We first investigate J_1. Using the fundamental property of the K.-B. functions we find

$$|J_1| \leqslant \frac{\tau+\delta}{\delta}\,\phi\left(\frac{\delta_s}{\tau+\delta}\right)$$

Let I* be any bounded, closed, ε independent interval

$$I^* = \{\tau \mid 0 \leqslant \tau \leqslant A^*\}$$

$$\|J_1\|_{I^*} \leqslant \max_{\tau\in I^*} \left\{\frac{\tau+\delta}{\delta}\,\phi\left(\frac{\delta_s}{\tau+\delta}\right)\right\} \leqslant \frac{A^*+\delta}{\delta}\,\phi\left(\frac{\delta_s}{\delta}\right)$$

If now

$$\frac{\delta_s}{\delta} = o(1), \text{ then}$$

$$\delta_1(\varepsilon) = \phi\left(\frac{\delta_s(\varepsilon)}{\delta(\varepsilon)}\right) = o(1)$$

Hence

$$J_1 = O\left(\frac{\delta_1}{\delta}\right) + O(\delta_1)$$

We next investigate J_2. The analysis is somewhat more delicate then in the case of J_1, because the fundamental estimate of K.-B. functions cannot be applied for all $\tau \in I^*$. We therefore subdevide I* as follows :

1) $0 \leqslant \tau \leqslant \delta_s$: Elementary estimation shows

$$J_2 = O\left(\frac{\delta_s}{\delta}\right)$$

2) $\delta_s \leqslant \tau \leqslant \delta'$ where δ' is an order function such that

$$\frac{\delta_s}{\delta'} = o(1) \ \underline{\text{and}}\ \frac{\delta'}{\delta} = o(1).$$

Again using elementary estimation we have

$$|J_2| \leqslant \frac{\delta_s}{\delta}\int_0^{\frac{1}{\delta_s}\tau} |\underline{\Phi}(t,\varepsilon) - \Phi_0|\,dt \Rightarrow J_2 = O\left(\frac{\delta'}{\delta}\right)$$

3) $\delta' \leqslant \tau \leqslant \delta$, with δ' defined as before. We now can use the fundamental property of K.-B. functions and obtain

$$|J_2| \leqslant \frac{\tau}{\delta}\,\phi\left(\frac{\delta_s}{\tau}\right) \leqslant \phi\left(\frac{\delta_s}{\delta'}\right)$$

This again implies $J_2 = o(1)$.

4) $\delta \leqslant \tau \leqslant A^*$. Here the estimation proceeds as in the case of J_1.

$$|J_2| \leqslant \frac{\tau}{\delta}\,\phi\left(\frac{\delta_s}{\tau}\right) \leqslant \frac{A^*}{\delta}\,\phi\left(\frac{\delta_s}{\delta}\right) \Rightarrow J_2 = O\left(\frac{\delta_1}{\delta}\right)$$

Hence, under the conditions specified in Lemma 6 we have

$$J_1 = o(1) \text{ and } J_2 = o(1)$$

which proofs the Lemma.

4. THE FUNDAMENTAL THEOREM.

We now investigate local averages of the function $\underline{Y}^*(\tau,\varepsilon)$ defined as solution of

$$\frac{d\underline{Y}^*}{d\tau} = \underline{F}(\underline{Y}^*, \frac{\tau}{\varepsilon}, \varepsilon) \; ; \; \underline{Y}^*(0,\varepsilon) = \underline{Y}_0$$

We have

$$\underline{Y}^*(\tau,\varepsilon) = \underline{Y}_0 + \int_0^{\tau} \underline{F}[\underline{Y}^*(\tau'',\varepsilon), \frac{\tau''}{\varepsilon}, \varepsilon]\, d\tau''$$

Using definition 2 we find for the local average :

$$\overline{\underline{Y}}(\tau,\varepsilon) = \underline{Y}_0 + \int_0^1 \{ \int_0^{\tau+\delta\overline{\tau}} \underline{F}[\underline{Y}^*(\tau'',\varepsilon), \frac{\tau''}{\varepsilon}, \varepsilon]\, d\tau''\} d\overline{\tau}$$

We shall deduce from this expression a relation for $\overline{\underline{Y}}(\tau,\varepsilon)$, not containing $\underline{Y}^*(\tau,\varepsilon)$.
We rewrite the right-hand side as follows

$$\overline{\underline{Y}}(\tau,\varepsilon) = \underline{Y}_0 + \int_0^1 \{ \int_{\delta\overline{\tau}}^{\tau+\delta\overline{\tau}} \underline{F}[\underline{Y}^*(\tau'',\varepsilon), \frac{\tau''}{\varepsilon}, \varepsilon] d\tau''\} d\overline{\tau} + I_1$$

where

$$I_1 = \int_0^1 \int_0^{\delta\overline{\tau}} \underline{F}[\underline{Y}^*(\tau'',\varepsilon), \frac{\tau''}{\varepsilon}, \varepsilon] d\tau'' d\overline{\tau}$$

It is immediately obvious that

$$|I_1| \leqslant \tfrac{1}{2} M\delta$$

In the remaining integral on the right-hand side of the expression for $\overline{\underline{Y}}(\tau,\varepsilon)$ we introduce a change of the integration variable

$$\tau'' = \tau' + \delta\overline{\tau}$$

and we subsequently interchange the order of integration. It follows that

$$\overline{\underline{Y}}(\tau,\varepsilon) = \underline{Y}_0 + \int_0^{\tau}\{\int_0^1 \underline{F}[Y^*(\tau'+\delta\overline{\tau},\varepsilon), \frac{\tau'}{\varepsilon} + \frac{\delta}{\varepsilon}\overline{\tau}, \varepsilon]\, d\overline{\tau}\} d\tau' + I_1$$

Finally, we write

$$\overline{\underline{Y}}(\tau,\varepsilon) = \underline{Y}_0 + \int_0^{\tau}\{\int_0^1 \underline{F}[\overline{\underline{Y}}(\tau',\varepsilon), \frac{\tau'}{\varepsilon} + \frac{\delta}{\varepsilon}\overline{\tau}, \varepsilon] d\overline{\tau}\} d\tau' + I_1 + I_2$$

where

$$I_2 = \int_0^{\tau} \int_0^1 \{\underline{F}[\underline{Y}^*(\tau'+\delta\overline{\tau},\varepsilon), \frac{\tau'}{\varepsilon} + \frac{\delta}{\varepsilon}\overline{\tau}, \varepsilon] - \underline{F}[\overline{\underline{Y}}(\tau',\varepsilon), \frac{\tau'}{\varepsilon} + \frac{\delta}{\varepsilon}\overline{\tau}, \varepsilon]\} d\overline{\tau}\, d\tau'$$

In order to estimate I_2 we use the Lipschitz-continuity of $\underline{F}$ and obtain

$$|I_2| \leqslant \lambda \int_0^{\tau} \int_0^1 |\underline{Y}^*(\tau'+\delta\overline{\tau},\varepsilon) - \overline{\underline{Y}}(\tau',\varepsilon)|\, d\overline{\tau}\, d\tau'$$

Using now Lemma's 4 and 5, we have

$|I_2| \leqslant \lambda M \delta \tau$

The above results are summarised in :

Lemma 7. If :

$$Y^*(\tau,\varepsilon) = Y_0 + \int_0^{\tau} \underline{F}[\underline{Y}^*(\tau',\varepsilon),\frac{\tau'}{\varepsilon},\varepsilon\ d\tau'$$

then

$$\overline{\underline{Y}}(\tau,\varepsilon) = \underline{Y}_0 + \int_0^{\tau}\int_0^1 \underline{F}[\overline{\underline{Y}}(\tau',\varepsilon),\frac{\tau'}{\varepsilon}+\frac{\delta}{\varepsilon}\overline{\tau},\varepsilon]\,d\overline{\tau}\ d\tau' + I_1 + I_2$$

where $|I_1| \leqslant \frac{1}{2}M\delta$, $|I_2| \leqslant \lambda M\delta\tau$

Using now theorem I and Lemma 4, we obtain

Lemma 8. If

$$Y^*(\tau,\varepsilon) = Y_0 + \int_0^{\tau} \underline{F}[\underline{Y}^*(\tau,\varepsilon),\frac{\tau}{\varepsilon},\varepsilon]\,d\tau'$$

then the local average $\overline{\underline{Y}}(\tau,\varepsilon)$ of $\underline{Y}^*(\tau,\varepsilon)$ can be approximated by the function $\tilde{Y}(\tau,\varepsilon)$, satisfying

$$\tilde{\underline{Y}}(\tau,\varepsilon) = \underline{Y}_0 + \int_0^{\tau}\int_0^1 F[\tilde{\underline{Y}}(\tau',\varepsilon),\frac{\tau'}{\varepsilon}+\frac{\delta}{\varepsilon}\overline{\tau},\varepsilon]\,d\overline{\tau}\ d\tau'$$

we have

$$\overline{\underline{Y}}(\tau,\varepsilon) = \tilde{\underline{Y}}(\tau,\varepsilon) + O(\delta)$$

and

$$\underline{Y}^*(\tau,\varepsilon) = \overline{\underline{Y}}(\tau,\varepsilon) + O(\delta)$$

the extimates being valid on any closed interval on which $\tilde{Y}(\tau,\varepsilon)$ exists and $\tilde{Y} \in D_0$.

Lemma 8 is the general and fundamental result, permitting to approximate $\underline{Y}^*(\tau,\varepsilon)$ by a function which is an approximation of the local average of $\underline{Y}^*(\tau,\varepsilon)$. In the case in which $F(\underline{Y},t,\varepsilon)$ is a K.-B. function lemma 8 reduces to the well-known fundamental theorem of the asymptotic theory of standard systems. We then have :

Theorem II.

Let $\underline{Y}^*(\tau,\varepsilon)$ be the solution of $\frac{d\underline{Y}^*}{d\tau} = F(\underline{Y}^*,\frac{\tau}{\varepsilon},\varepsilon)$; $\underline{Y}^*(0,\varepsilon) = \underline{Y}_0 \in D_0$ and let $\underline{\eta}(\tau)$ be the solution of

$$\frac{d\eta}{d\tau} = \underline{F}_0(\underline{\eta})\ ;\quad \underline{\eta}(0) = \underline{Y}_0$$

where

$$\underline{F}_0(\underline{\eta}) = \lim_{T\to\infty} \frac{1}{T}\int_0^T \underline{F}(\underline{\eta},t,0)\,dt$$

Suppose $\underline{\eta}(\tau)$ exists for $0 \leqslant \tau \leqslant A$, and $\underline{\eta} \in D_0$, then $\underline{Y}^*(\tau,\varepsilon)$ exists in the same interval and

$$\underline{Y}^*(\tau,\varepsilon) = \underline{\eta}(\tau) + o(1)$$

Proof.

Because of the uniform behaviour of $\underline{F}(\underline{Y},t,\varepsilon)$ as $\varepsilon \downarrow 0$, there exists an order function $\delta_2(\varepsilon) = o(1)$ such that for all $\underline{Y} \in D$ and $0 \leqslant t < \infty$:

$$|\underline{F}(\underline{Y},t,\varepsilon) - \underline{F}(\underline{Y},t,0)| \leqslant \delta_2(\varepsilon) \ ; \ \delta_2(\varepsilon) = o(1)$$

Because $\underline{F}(\underline{Y},t,\varepsilon)$ is a K.-B. function, we may use the result of Lemma 6. It follows that

$$\underline{\tilde{Y}}(\tau,\varepsilon) = \underline{Y}_0 + \int_0^\tau F_0(\underline{\tilde{Y}}(\tau'))d\tau' + I_3$$

where

$$|I_3| \leqslant M'(\frac{\delta_1}{\delta} + \delta_2)\tau$$

and M' is a constant independent of ε.

Using now theorem I we find

$$|\ \underline{Y}(\tau,\varepsilon) - \underline{\eta}(\tau)| \leqslant \frac{1}{\lambda}M'(\frac{\delta_1}{\delta} + \delta_2)\ (e^{\lambda\tau} - 1)$$

Hence, for any interval $0 \leqslant \tau \leqslant A$, for which $\eta(\tau)$ exists, $\underline{\tilde{Y}}(\tau,\varepsilon)$ exists and

$$\underline{\tilde{Y}}(\tau,\varepsilon) = \underline{\eta}\ (\tau) + O(\frac{\delta_1}{\delta}) + O(\delta_2)$$

From Lemma 8 it follows now that

$$\underline{Y}^*\ (\tau,\varepsilon) = \underline{\eta}(\tau) + O(\frac{\delta_1}{\delta}) + O(\delta) + O(\delta_2)$$

Since $\delta(\varepsilon) = o(1)$ is an order function such that $\frac{\delta_1}{\delta} = o(1)$, $\underline{\eta}(\tau)$ indeed is an asymptotic approximation of $\underline{Y}^*(\tau,\varepsilon)$, that is

$$\underline{Y}\ (\tau,\varepsilon) = \underline{\eta}(\tau) + o(1).$$

5. FURTHER RESULTS.

In the proceeding section we have established the fundamental theorem of Krilov-Bogolioubov under the most general conditions. Further improved results for periodic systems (that is systems in which $F(\underline{Y},t,\varepsilon)$ is a finite some of functions periodic in the t variable), and for slowly varying periodic systems in the sense of Mitropolski, can also easily be deduced by the present method of analysis. Furthermore, higher approximations for such systems can be constructed. This is shown in Eckhaus (1974). These results are well-known, but they are reproduced by a relatively simple deductive analysis.

However, the main object of our investigation was not to rederive, in what is hoped to be a more satisfactory way, the classical results of the asymptotic theory of non-linear oscillations. Our aim was to develop a method of analysis which would permit to derive new results. These are concerned with enlarging the domain of vality of the asymptotic approximation to the whole time axis, and with applications to problems in partial differential equations.We summarize here some of the results obtained in Eckhaus (1974).

In the asymptotic theory of periodic systems the following result is well-known (see for example Roseau (1966)) :

Suppose the associated system

$$\frac{d\underline{\eta}}{d\tau} = \underline{F}_0(\underline{\eta}) \; ; \; \tau = \varepsilon t$$

has a singular point $\underline{\xi}$ which is asymptotically stable in linear approximations. Then there exists a periodic solution $\underline{\tilde{Y}}(t,\varepsilon)$ such that

$$\lim_{\varepsilon \to 0} |\underline{\tilde{Y}}(t,\varepsilon) - \underline{\xi}| = 0$$

uniformly in t.
In Eckhaus (1974) we prove, under similar conditions :

Theorem II.

Let $\underline{Y}(t,\varepsilon)$ be the solution of the periodic system

$$\frac{d\underline{Y}}{dt} = \varepsilon F(\underline{Y},t) \; ; \; \underline{Y}(0,\varepsilon) = \underline{Y}_0 \in D_0 .$$

Let $\underline{\eta}(\tau)$, with $\tau = \varepsilon t$, be the solution of the associated system

$$\frac{d\underline{\eta}}{d} = \underline{F}_0(\underline{\eta}) \; ; \; \underline{\eta}(0) = \underline{Y}_0$$

where $\underline{F}_0(\underline{\eta}) = \lim_{T\to\infty} \frac{1}{T} \int_0^T F(\underline{\eta},t)dt.$

Suppose that :

i. $\underline{\eta} = \underline{\xi}$ is a singular point of the associated system ;
$\underline{\eta} = \underline{\xi}$ is asymptotically stable in the linear approximation.

ii. $\underline{Y}_0$ belongs to the domain of attraction of $\underline{\xi}$.

iii. $\underline{\eta}(\tau) \in D_0$ for $0 \leqslant \tau < \infty$.

Then :

$$\underline{Y}(t,\varepsilon) = \underline{\eta}(\varepsilon t) + o(1)$$

uniformly on $0 \leqslant t < \infty$.

Finally, it is tempting to investigate whether the concept of asymptotic average values could also be useful in studying problems gouverned by partial differential equations of the type describing wave-propagation phenomena. For such problems various formal methods have been proposed, however the methods generally do not contain proof of the asymptotic validity of the results.
We have studied a class of problems investigated by Chikwendu and Kevorkian (1972). These authors have proposed a formal method of construction of asymptotic approximations (without proof of the validity of the result). Using now local average values, the fundamental result of Chikwendu and Kevorkian and conditions for validity of the asymptotic approximation can be derived.
Thus, for the class of problems of Chikwendu and Kevorkian, that is for the perturbed wave equation

$$\frac{\partial^2 u}{\partial t^2} - \frac{\partial^2 u}{\partial x^2} = \varepsilon \, H(\frac{\partial u}{\partial t}, \frac{\partial u}{\partial x})$$

our approach permits to develop a deductive theory containing proof of the asymptotic validity of the results.

It seems reasonable to expect that the example of the perturbed wave equation indicates further possibilities in using the concept of asymptotic local average values (as defined here in section 3) for the study of wave propagation phenomena.

REFERENCES.

Bogoliubov, N.N and Mitropolski, Y.A.
"Asymptotic methods in the theory of non-linear oscillations".
Gordon & Breach, New York. (1961).

Chikwendu, S.C. and Kevorkian, J.
"A perturbation method for hyperbolic equations with small non-linearities".
SIAM Journal on Applied Mathematics, vol. 27. no. 2 (1972).

Coddington, E.A. and Levinson, N.
"Theory of ordinary differential equations".
McGraw-Hill, New York (1955)

Eckhaus, W "New approach to the asymptotic theory of non-linear oscillations and wave propagation".
to appear in Journ. of Math. Analysis and Applications (1974).

Mitropolski, Y.A.
"Problems de la théorie asymptotique des oscillations non-stationnaires".
Gauthier-Villars Edition (1966).

Roseau, M
"Vibrations non-linéaires et théorie de la stabilité".
Springer-Verlag (1966).

Volosov, V.M.
"Averaging in systems of ordinary differential equations".
Russ. Math. Surveys 17, 6. (1962).

ON FIRST ORDER MATCHING PROCESS FOR SINGULAR FUNCTIONS

J. MAUSS
Département de Mathématiques. Université Paul SABATIER.
Toulouse. FRANCE

In this paper, we consider the techniques of matching which have been proposed to yield a relationship between inner and outer expansion of a singular function. The method of matched asymptotic approximation is well known and many problems, especially in fluid Mechanics, have been successfully treated using it. However, the method has no actual justification and there is no rigorous mathematical proof for it. The first who worked about the foundation of the matching procedures were S. KAPLUN [1] with the so-called extension theorem. Nevertheless, there were not a lot of developments from Kaplun's ideas till W. ECKHAUS [2] tried to attempt a systematic approach of the theory of matching. In practice, as long as we are interested in singular perturbations problems, it is quite useful to get matching rules in a simple way. As L.E. FRAENKEL [3] stated it, the techniques which use the idea of over-lapping, with the concepts of Kaplun and Lagerstrom [1], is often difficult and laborious. Later, M. VAN DYKE [4] thought to express the mind of Kaplun when stating his rule ; this matching principle is very simple in application, but unfortunatly, it is not exactly true.

Here, using the ideas of ECKHAUS, we try to show how Van Dyke's matching rule could be the best one. In a first step, we recall, some asymptotic definitions ; an extensive study on these points can be found in Eckhaus's book [5]. Secondly, extension theorem in the form stated in [2] will be considered. Finally, matching principle and uniform approximation of singular functions are studied by using some simple examples.

1. ASYMPTOTIC DEFINITIONS

Let $\varphi(x,\varepsilon)$ be a function of the real variable x and real parameter ε, defined in a bounded closed domain $\mathcal{D}$: $0 \leqslant x \leqslant B_0$ and $0 \leqslant \varepsilon \leqslant \varepsilon_0$, where B_0 and ε_0 are positive constants. Moreover, $\varphi(x,\varepsilon)$ is supposed to be continuous both in x and ε.

Consider now the set $\mathcal{E}$ of order functions $\delta(\varepsilon)$; they are real, positive and continuous in $0 < \varepsilon \leqslant \varepsilon_0$ such that $\lim_{\varepsilon\to 0} \delta(\varepsilon)$ exists.

Also, we introduce the subset $\bar{\mathcal{E}} \subset \mathcal{E}$ such that if $\delta \in \bar{\mathcal{E}}$, then, $\lim_{\varepsilon\to 0} \delta(\varepsilon) = 0$.

In the following analysis, we use the norm of uniform convergence,

$$\|\varphi\| = \underset{\mathcal{D}\,:\, 0\leqslant x\leqslant B_0}{\text{Max}} |\varphi(x,\varepsilon)|$$

and the notations,

$\varphi = O(\delta)$ if $\|\varphi\| < K\delta$ where K is a constant ;

$\varphi = o(\delta)$ if $\lim_{\varepsilon\to 0} \dfrac{\|\varphi\|}{\delta} = 0$.

In comparing two order function δ_1 and δ_2, we shall use also,

$\delta_1 \simeq \delta_2$ if $\lim_{\varepsilon\to 0} \dfrac{\delta_1}{\delta_2} = 1$,

$\delta_1 << \delta_2$ if $\delta_1 = o(\delta_2)$.

A regular asymptotic expansion of the function $\varphi(x,\varepsilon)$ is defined by,

$$(1) \quad \varphi(x,\varepsilon) = E_0^{(m)}\varphi + o(\delta_0^{(m)}),$$

where $E_0^{(m)}$ is an expansion operator [3, 5] to the order m such that,

$$E_0^{(m)}\varphi = \sum_{n=0}^{m} \delta_0^{(n)}(\varepsilon)\, \varphi_0^{(n)}(x) \ ;$$

the sequence $\delta_0^{(n)}$ fulfilled the following asymptotic inequality,

$$\delta_0^{(n+1)} << \delta_0^{(n)} \text{ for all } n.$$

Finally, functions $\varphi_0^{(n)}(x)$ are obtained by limit processes,

$$(2) \quad \varphi_0^{(n)}(x) = \lim_{\varepsilon \to 0} \frac{1}{\delta_0^{(n)}(\varepsilon)} \left[\varphi - E_0^{(n-1)} \varphi\right].$$

It can be shown that if an asymptotic expansion exists in a subdomain $\bar{\mathcal{D}} \subset \mathcal{D}$, then the function $\varphi_0^{(n)}(x)$ are continuous functions of x in $\bar{\mathcal{D}}$ [2].

If such a limit process is non uniform, for instance to the first order, in the whole domain $\mathcal{D}$, the function φ is said singular. The asymptotic approximation of a singular function will be our task in the following. Moreover, we study the case in which non uniformity of φ occurs at isolated points of $\mathcal{D}$ and precisely at $x = 0$. Thus, $E_0^{(o)}\varphi$ is an asymptotic approximation of φ in $A_0 \leqslant x \leqslant B_0$, where A_0 is a strictly positive constant, but is not in the neighborhood of $x = 0$.

It is very usual in Mathematical Physics to call expansion (1) outer expansion and, more precisely, $E_0^{(o)}\varphi$ is the outer approximation.

To study the neighborhood of the origin, local variables are introduced :

$$x_\nu = \frac{x}{\delta_\nu(\varepsilon)} \quad \text{where} \quad \delta_\nu \in \bar{\mathcal{E}}.$$

After this stretching transformation, we can define local regular expansions in $\mathcal{D}_\nu$: $A_\nu \leqslant x_\nu \leqslant B_\nu$ where A_ν and B_ν are two positive constants,

$$\varphi(x,\varepsilon) = E_\nu^{(m)}\varphi + o(\delta_\nu^{(m)}),$$

$$E_\nu^{(m)}\varphi = \sum_{n=0}^{m} \delta_\nu^{(n)}(\varepsilon)\, \varphi_\nu^{(n)}(x_\nu).$$

the expansion operator $E_\nu^{(m)}$ and functions $\varphi_\nu^{(n)}$ are defined exactly in the same way than (1) and (2).

Here, we use the norm definition,

$$\|\varphi\|_\nu = \underset{\mathcal{D}_\nu : A_\nu \leqslant x_\nu \leqslant B_\nu}{\text{Max}} |\varphi(x_\nu \delta_\nu, \varepsilon)| ,$$

such that, in $\mathcal{D}_\nu$,

$$\varphi = o(\delta) \text{ if } \lim_{\varepsilon\to 0} \frac{\|\varphi\|_\nu}{\delta} = 0.$$

The process which consists to relate functions $E_\nu^{(m)}\varphi$ to each others is called matching. This process can take various forms, one of them is the so-called extension theorem [1]. By hypothesis, $\varphi(x,\varepsilon)$ converges uniformly, as $\varepsilon \to 0$, on $A_0 \leqslant x \leqslant B_0$, but the convergence is not uniform on $0 \leqslant x \leqslant B_0$. The extension theorem asserts that the domain of uniform convergence of φ can, in a sense, be extended to include the origin.

We now state an important theorem the proof of which can be found in [2] ; this theorem is a consequence of the extension theorem.

Theorem. Let $E_{\nu_1}^{(o)}\varphi$ and $E_{\nu_2}^{(o)}\varphi$ be two local asymptotic approximations,

$$E_{\nu_1}^{(o)}\varphi = \delta_{\nu_1}^{(o)} \; \varphi_{\nu_1}^{(o)}(x_{\nu_1}) \; , \; E_{\nu_2}^{(o)}\varphi = \delta_{\nu_2}^{(o)} \; \varphi_{\nu_2}^{(o)}(x_{\nu_2}) \text{ with}$$

$$x_{\nu_1} = \frac{x}{\delta_{\nu_1}} \; , \; x_{\nu_2} = \frac{x}{\delta_{\nu_2}} \quad \text{and} \quad \delta_{\nu_2} \ll \delta_{\nu_1}.$$

There exists an order function $\delta_\nu^{(1,2)} \ll 1$ such that if

$$\delta_\nu^{(1,2)} \ll \frac{\delta_{\nu_2}}{\delta_{\nu_1}} \ll 1 \; ,$$

then, for $\delta_\mu : \delta_{\nu_2} \ll \delta_\mu \ll \delta_{\nu_1}$, we have,

$$E_\mu^{(o)}E_{\nu_1}^{(o)}\varphi = E_\mu^{(o)}\varphi = E_\mu^{(o)}E_{\nu_2}^{(o)}\varphi \quad \text{with } E_\mu^{(o)}\varphi = \delta_\mu^{(o)}\varphi_\mu^{(o)}(x_\mu).$$

This theorem is very important ; it is the basis of the technique of matching with the idea of over_lapping, using intermediate expansions. But one must be very cautious when using such a theorem. In practice, the subset of $\bar{\mathcal{E}}$ choosen for the stretching order functions is not large enough to apply the theorem. Thus, it will be important to get a rule, as Van Dyke's one, independantly of this choice. For instance, if we consider the function,

$$\varphi(x,\varepsilon) = \frac{1}{\operatorname{Log} x} + \frac{e^{-\frac{x}{\varepsilon}}}{\operatorname{Log} \varepsilon},$$

with $x_\nu = \frac{x}{\varepsilon^\nu}$ $(\nu > o)$, we obtain,

$$E_o^{(o)}\varphi = \frac{1}{\operatorname{Log} x}, \quad E_\nu^{(o)}\varphi = \frac{1}{\nu \operatorname{Log} \varepsilon}, \quad E_1^{(o)}\varphi = \frac{1+e^{-x_1}}{\operatorname{Log} \varepsilon}.$$

Now, the theorem asserts there exists $\mu < 1$ such that

$$\frac{1}{\operatorname{Log} \varepsilon} = E_\mu^{(o)} E_1^{(o)}\varphi = E_\mu^{(o)}\varphi = \frac{1}{\mu \operatorname{Log} \varepsilon},$$

so, we get $\mu = 1$ which contradicts the hypothesis. We shall consider this example later.

2. MATCHING PRINCIPLE

In the same way as the set ε^ν, the set $\delta_\nu(\varepsilon)$ is ordered such that for $\nu_2 > \nu_1$, $\delta_{\nu_2} << \delta_{\nu_1}$ and $\delta_o \simeq 1$.
Moreover, we assume that for $\nu < 1$, there exists m_ν such that,

$$E_\nu^{(o)}\varphi = E_\nu^{(o)} E_o^{(m_\nu)}\varphi,$$

but, for $\nu = 1$, it is not possible to find such an m_1. This means that if for $\nu < 1$, intermediate approximation $E_\nu^{(o)}\varphi$ are contained in the outer expansion, it is not the case for $E_1^{(o)}\varphi$; we say that $E_1^{(o)}\varphi$ is relatively significant. This terminology is used to distinguish with the definition of a significant approximation in [5].

Now, we will establish a matching principle between $E_o^{(o)}\varphi$ and $E_1^{(o)}\varphi$ and thus, construct a uniform asymptotic approximation in $A_1\delta_1(\varepsilon) \leqslant x \leqslant B_o$. Of course, it is supposed that $m_\nu = 0$ but this is not a hard restriction since we can study the function $\varphi - E_o^{(m_\nu - 1)}\varphi$ instead of φ.

Moreover, if $A_1 = 0$, $E_1^{(o)}\varphi$ is the first term of the inner expansion, the inner approximation. If not, $E_1^{(o)}\varphi$ is the first approximation of an intermediate "boundary layer" ; but since the matching has to be done between consecutive relatively significant approximations, we are not studying the case of multiple boundary layers in the vicinity of $x = 0$. Thus, for simplicity, even if A_1 is not zero, we shall call $E_1^{(o)}\varphi$ the inner approximation.

Now we state a theorem which is a matching principle with a rule analogous as Van Dyke's one.

In the following, all superscripts (0) in the expansion operators will be omitted.

Theorem 2. If for $0 \leqslant \nu < 1$, the following hypothesis is satisfied, (I) $E_\nu E_0 \varphi = E_\nu \varphi$

with additionnal condition,

$$\|E_0\varphi\|_1 \simeq \|\varphi\|_1 ,$$

then for every ν such that $0 \leqslant \nu < 1$, we get,

(3) $E_\nu E_1 E_0 \varphi = E_\nu E_1 \varphi$, particularly,

(3)' $E_0 E_1 E_0 \varphi = E_0 E_1 \varphi$.

Since from Eckhaus [2], there is continuity of the set of order functions $\|\varphi\|_\nu$ with respect to the set δ_ν, the additional condition is always satisfied when condition (I) is fulfilled. Nevertheless, in practice, when particular subsets for δ_ν are used, this condition must be verified ; e.g the function $\varphi = x + e^{-\frac{x}{\varepsilon}}$ with the subset $\delta_\nu = \varepsilon^\nu$ is such that condition (I) is satisfied but, $\|E_0\varphi\|_1 \simeq \varepsilon$ while $\|\varphi\|_1 \simeq 1$. For the proof of this theorem, the following lemma will be needed :

Lemma. If Δ_1 and Δ_2 are two order functions such that

$$\Delta_2 = o(\Delta_1) ,$$

then, for all $\lambda > 0$ and $\delta \in \overline{\mathcal{E}}$, we have,

$$\Delta_2(\lambda\delta) = o[\Delta_1(\lambda\delta)].$$

Now, with theorem 1, there exists μ_1, such that for $\mu_1 \leqslant \nu < 1$,

(4) $E_\nu \varphi = E_\nu E_1 \varphi$.

By using hypothesis (I),

(5) $E_\nu E_0 \varphi = E_\nu E_1 \varphi$.

This last rule is well known (intermediate matching) but it is only exact for ν in the neighborhood of 1. Thus, if the set δ_ν is a subset of $\bar{\mathcal{E}}$, it happens that this equality fails and it seems there is no matching at all. If now we consider function $E_0\varphi$, there exists μ_2 such that if $\mu_2 \leq \nu < 1$, we have by (4),

$$(6)\quad E_\nu E_0\varphi = E_\nu E_1 E_0\varphi \ ;$$

From (5) and (6) we obtain

$$(3)\quad E_\nu E_1 E_0\varphi = E_\nu E_1\varphi \ .$$

In fact, there exists $\mu = \max(\mu_1,\mu_2)$ such that if $\mu \leq \nu < 1$, this last equality is true with $E_\nu\varphi$ as common value. We intend to show this equality is still exact when $\nu < \mu$; of course the common value is no longer $E_\nu\varphi$.

We only outline the proof since the lemma is the essential point. Let us suppose that the asymptotic value of $E_1\varphi$ when $x_1 \nearrow \infty$ is given by

$$(7)\quad E_1\varphi = a_1\Delta_1(\frac{1}{x_1}) + o(\Delta_1) \text{ with } \Delta_1 \in \mathcal{E} \text{ and where } a_1 \text{ is}$$

a constant. Then, by the lemma,

$$\|E_1\varphi\|_\nu \simeq a_1\ \Delta_1\ (\frac{1}{x_\nu}\ \frac{\delta_1}{\delta_\nu}).$$

Moreover, the asymptotic behaviour of function $E_1\varphi - E_1E_0\varphi$, for $x_1 \nearrow \infty$, is such that

$$E_1\varphi - E_1E_0\varphi = a_2\Delta_2(\frac{1}{x_1}) + o(\Delta_2),$$

where a_2 is a constant ; again, $\Delta_2 \in \mathcal{E}$ and

$$\|E_1\varphi - E_1E_0\varphi\|_\nu \simeq a_2\Delta_2(\frac{1}{x_\nu}\quad \frac{\delta_1}{\delta_\nu})$$

But now, by (3), we know that $\Delta_2 = o(\Delta_1)$.

Then, using the lemma once more, this asymptotic inequality is still true for all x_ν and δ_ν so that we get,

$$\|E_1\varphi - E_1E_0\varphi\|_\nu = o(\|E_1\varphi\|_\nu) \text{ for } 0 \leq \nu < 1.$$

It will be the same kind of proof if the asymptotic nature of function $E_1\varphi$ in (7) were given by $\dfrac{a_1}{\Delta_1(\frac{1}{x_1})}$ for $x_1 \nearrow \infty$.

Conversely, it is easy to see that we obtain analogous results if $E_\nu\varphi$ is contained, for $0 < \nu \leq 1$ in the inner approximation.

Theorem 3. If for $0 < \nu \leq 1$, the following hypothesis is satisfied, (II) $E_\nu E_1 \varphi = E_\nu \varphi$,
with additional condition,

$$\|E_1 \varphi\|_0 \simeq \|\varphi\|_0 ,$$

then, for every ν such that $0 < \nu \leq 1$, we get,

(8) $E_\nu E_0 E_1 \varphi = E_\nu E_0 \varphi$, particularly,

(8)' $E_1 E_0 E_1 \varphi = E_1 E_0 \varphi$.

Thus, these two theorems provide us a matching treatment ; firstly, intermediate matching which is now systematic, secondly, rules as Van Dyke's one which are very easy to apply. It is simple to construct now a uniform asymptotic approximation valid for $A_1 \delta_1(\varepsilon) \leq x \leq B_0$.

Corollary. If φ_0 and ψ_0 are defined in the following way,

(9) $\varphi_0 = E_0 \varphi + E_1 \varphi - E_1 E_0 \varphi$,

(10) $\psi_0 = E_0 \varphi + E_1 \varphi - E_0 E_1 \varphi$,

then φ_0 (resp ψ_0) is a uniform asymptotic approximation of φ in $A_1 \delta_1(\varepsilon) \leq x \leq B_0$ if condition (I) (resp. condition (II)) is satisfied.

As Eckhaus [4] and Fraenkel [3] noted it, when Van Dyke's rule is to be applied, there are two available forms for the matching and thus, to construct uniform approximations.

Let us now take some simple examples.

1°) $\varphi(x,\varepsilon) = \dfrac{1}{\operatorname{Log} x} + \dfrac{e^{-\frac{x}{\varepsilon}}}{\operatorname{Log} \varepsilon}$.

This example has been considered yet and we have seen the set ε^ν is not sufficient to get intermediate matching.

If we put, $x_\nu = \dfrac{x}{\varepsilon^\nu}$, we obtain,

$$E_0 \varphi = \frac{1}{\operatorname{Log} x} , \quad E_\nu \varphi = \frac{1}{\nu \operatorname{Log} \varepsilon} , \quad E_1 \varphi = \frac{1+e^{-x_1}}{\operatorname{Log} \varepsilon} .$$

Here, condition (I) is satisfied, so is the additionnal condition in such a way theorem 2 gives us,

(3) $\dfrac{1}{\operatorname{Log} \varepsilon} = E_\nu E_1 E_0 \varphi = E_\nu E_1 \varphi = \dfrac{1}{\operatorname{Log} \varepsilon}$ for $\nu < 1$,

but, $\dfrac{1}{\operatorname{Log} \varepsilon} = E_\nu E_0 E_1 \varphi \neq E_\nu E_0 \varphi = \dfrac{1}{\nu \operatorname{Log} \varepsilon}$.

2°) $\varphi(x,\varepsilon) = \dfrac{1}{\text{Log } x - \text{Log } \varepsilon + 1}$

In this example, always with $x_\nu = \dfrac{x}{\varepsilon^\nu}$, theorem 3 is to be applied.

$$E_0\varphi = -\frac{1}{\text{Log } \varepsilon},\quad E_\nu\varphi = \frac{1}{(\nu-1)\text{ Log } \varepsilon},\quad E_1\varphi = \frac{1}{\text{Log } x_1 + 1}$$

In fact, we have,

$$(8)\quad -\frac{1}{\text{Log } \varepsilon} = E_\nu E_0 E_1 \varphi = E_\nu E_0 \varphi = -\frac{1}{\text{Log } \varepsilon} \text{ for } \nu > 0,$$

3°) $\varphi(x,\varepsilon) = \dfrac{1}{\text{Log } x} + \dfrac{e^{-\frac{x}{\varepsilon}}}{\text{Log } \varepsilon} + \dfrac{1}{\text{Log } x - \text{Log } \varepsilon + 1}$

Here, the function considered is the sum of the last two functions. It is a very simple function but, there is no possible matching. Particularly, we have,

$$\frac{1}{\text{Log } \varepsilon} = E_\nu E_1 E_0 \varphi \neq E_\nu E_1 \varphi = \frac{1}{(\nu-1)\text{ Log } \varepsilon} \quad \text{for all } \nu\ ;$$

$$-\frac{1}{\text{Log } \varepsilon} = E_\nu E_0 E_1 \varphi \neq E_\nu E_0 \varphi = \frac{1}{\nu \text{ Log } \varepsilon} \quad \text{for all } \nu\ .$$

By this example, it is very clear that we have to use the method of multiple scales ; in fact, intermediate approximation is partly contained in the outer approximation, partly in the inner approximation :

$$E_0\varphi = \frac{1}{\text{Log } x},\quad E_\nu\varphi = \frac{1}{\text{Log } \varepsilon}\left[\frac{1}{\nu} + \frac{1}{\nu-1}\right],\quad E_1\varphi = \frac{1}{\text{Log } x_1 + 1}$$

4°) $\varphi(x,\varepsilon) = \dfrac{1}{\text{Log } x - \text{Log } \varepsilon + 1} + \dfrac{\text{Log }(x+\varepsilon)}{(\text{Log } \varepsilon)^2}$.

Here, we get an example where there is no intermediate matching and so, no uniform approximation but, Van Dyke's rule can be applied. That is because the second part of the function only appears in the intermediate approximation.

$$E_0\varphi = -\frac{1}{\text{Log } \varepsilon},\quad E_\nu\varphi = \frac{1}{\text{Log } \varepsilon}\left[\nu + \frac{1}{\nu-1}\right],\quad E_1\varphi = \frac{1}{\text{Log } x_1 + 1}$$

3. CONCLUSIONS.

In the last pages, we gave some indications about the

foundations of the method of Matched Asymptotic expansions ; particularly it was possible to state a rule which is very easy to apply and which is slightly different from Van Dyke's one, both by hypothesis and by conclusions. Nevertheless, if we consider once more the first two examples, we note that, in the two cases, we have for all ν,

$$(11) \quad E_\nu E_1 E_0 \varphi = E_\nu E_0 E_1 \varphi \, .$$

It was not possible to find any example where this equality was not fulfilled. We think it is not a coïncidence and that's why we will state a conjecture.

Conjecture. If for a given function $\varphi(x,\varepsilon)$, we can construct the function,

$$\psi = E_{\mu_2} E_{\mu_1} \varphi,$$

for μ_1 and μ_2 such that $\mu_2 \neq \mu_1$, then, for every couple ν_1, ν_2,

$$E_{\nu_2} E_{\nu_1} \psi = E_{\nu_2} \psi \, .$$

This conjecture arises from the fact that expansion operators are regularizing operators with respect to the matching. Applying the conjecture for instance for $\mu_2 = 0$ and $\mu_1 = 1$, we assert that for the function $\psi = E_0 E_1 \varphi$, the classical intermediate matching is correct on the whole range of values of ν :

$$(12) \quad E_\nu \psi = E_\nu E_1 \psi \; ;$$

we put there $\nu_1 = 1$ and $\nu_2 = \nu$.

Now, let us assume that hypothesis (II) is satisfied in theorem 3 ; we have,

$$(8)' \quad E_1 E_0 E_1 \varphi = E_1 E_0 \varphi \, ,$$

but, by (12), we obtain

$$E_\nu E_0 E_1 \varphi = E_\nu E_1 E_0 E_1 \varphi = E_\nu E_1 E_0 \varphi$$

which is (11). This relation holds for $\nu = 0$,

$$E_0 E_1 \varphi = E_0 E_1 E_0 \varphi \quad \text{which is (3)'.}$$

Thus, we have seen that hypothesis (II) leads to (3)' ; conversely, if hypothesis (I) is satisfied, then (8)' is too. So, if the conjecture is true and if one of the hypothesis (I) or (II) is satisfied, then, both of the forms of matching are equivalent. Moreover, we have (11) for all values of ν and it is easy to see that there is only one way, in the corollary, to construct a uniform approximation. That may be why there are not a lot of mistakes when Physicians or Mechanics used the method of matched asymptotic expansions.

UNIVERSITE PAUL SABATIER
118, Route de Narbonne
31077 TOULOUSE.

REFERENCES.

[1] KAPLUN. S. and LAGERSTROM.P.A (1957). Asymptotic expansions of Navier-Stokes solutions for small Reynolds numbers.
J. Math. and Mech, 6, 585.

[2] ECKHAUS. W. (1969). On the foundations of the method of matched asymptotic approximations. J de Mécanique, 8, 265.

[3] FRAENKEL. L. E. (1969). On the method of matched asymptotic expansions. Part. I : A matching principle. Proc. Camb. Phil. Soc, 65, 209.

[4] VAN DYKE. M. D. (1964). Perturbation methods in fluid mechanics. New-York. Academic Press.

[5] ECKHAUS . W. (1973). Matched Asymptotic expansions and singular perturbations. North-Holland. American Elsevier.

THE BIRTH OF A BOUNDARY LAYER IN AN ELLIPTIC SINGULAR PERTURBATION PROBLEM*)

J. GRASMAN
Mathematical Centre
Amsterdam, The Netherlands

Abstract: In elliptic singular perturbation problems two types of boundary layers may arise. In a particular example it is shown that there exists a smooth transition from one type of boundary layer to the other.

1. INTRODUCTION

We investigate the Dirichlet problem for the second order, linear, elliptic differential equation

$$L_\varepsilon \phi \equiv (\varepsilon L_2 + L_1)\phi = 0, \qquad 0 < \varepsilon << 1 \tag{1}$$

in a bounded convex domain G of $\mathbb{R}_2$. When L_1 is a first order differential operator with constant coefficients the solution of the problem exhibits a well-known boundary layer structure. There exists a large number of papers dealing with this subject, we mention Visik and Lyusternik (1962) and Eckhaus and De Jager (1966). In these studies it is remarked that some difficulties arise, if one tries to approximate ϕ in a neighborhood of a point where the subcharacteristic of L_1 is tangent to the boundary of G. In Grasman (1971) it is shown that the seemingly singular behaviour of the asymptotic approximation is due to the presence of corner layers, which are visualized by applying the coordinate stretching method to both coordinates. This technique has been suggested by Eckhaus (1968).

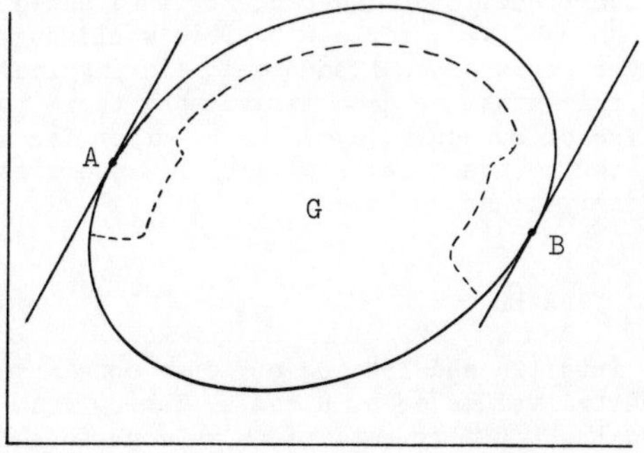

Fig. 1.

*) This paper is registered as Mathematical Centre Report TW 138/73.

Let us first consider the case where we have $(2n-1)^{th}$ order of tangency $(n=1,2,\ldots)$ of the subcharacteristics in the points A and B, which divide the boundary into two segments. It appears that then we have a boundary layer of thickness $O(\varepsilon)$ along one of these segments depending on the coefficients of L_1. This can be proved with the maximum principle for differential equations. Near A and B there are layers of thickness $O(\varepsilon^{\nu})$ and length $O(\varepsilon^{\mu})$, $\nu = 2n/(4n-1)$ and $\mu = 1/(4n-1)$. For $n \to \infty$ such a layer tends to a so-called parabolic boundary layer. In this paper we will attain this limit situation in a different way.

The differential equation which will be studied in particular is of the form

$$\varepsilon\Delta\phi + a\frac{\partial\phi}{\partial x} - b\frac{\partial\phi}{\partial y} - c\phi = 0 \qquad (a\geq 0,\ b>0). \tag{2}$$

For the domain $G = \{x,y;\ x \geq 0,\ y \geq 0\}$ the boundary values are

$$\phi(x,0) = 0 \qquad \text{for } x \geq 0, \tag{3a}$$

and

$$\phi(0,y) = f(y) \qquad \text{for } y > 0. \tag{3b}$$

Since the domain is not bounded, we only consider solutions that satisfy a boundedness condition at infinity. For the solution in rectangular domain we refer to Grasman (1974).

For $\varepsilon \to 0$ equation (2) degenerates to a first order differential equation. It can be demonstrated that the solution of this reduced equation only satisfies condition (3a). This trivial solution will hold in the greater part of the domain G, only a neighborhood of the line $x = 0$ needs to be excluded. Near this line ϕ has a boundary layer structure. Special attention will be given to the corner layer near the point (0,0). This is done by stretching both coordinates as follows

$$x = \xi\varepsilon^{\alpha}, \qquad y = \eta\varepsilon^{\beta}. \tag{4}$$

Near $x = 0$ we may expect an ordinary boundary layer of thickness $O(\varepsilon)$ for $a \geq \delta > 0$ (δ arbitrary small but independent of ε) and a parabolic boundary layer of thickness $O(\sqrt{\varepsilon})$, for $a = 0$. This would suggest that a parabolic boundary layer is an unusual phenomenon in physical problems.

It is the aim of this study to demonstrate that there is a smooth transition from one type of boundary layer to the other for $a \to 0$. In such a transition interval the boundary layer will have properties of both the parabolic and the ordinary boundary layer.

2. THE ORDINARY BOUNDARY LAYER

Substituting (4) into (2) and letting $\varepsilon \to 0$ we obtain various limiting equations depending on the values of α, β and a. Taking into account the matching principle and the boundary conditions we come to a certain number of significant asymptotic approximations satisfying some limiting equation.

For $a \geq \delta > 0$ we distinguish the following significant cases

(a) *The ordinary boundary layer*

For $\alpha = 1$ and $\beta = 0$ we have the limiting equation

$$\frac{\partial^2 U}{\partial\xi^2} + a\frac{\partial U}{\partial y} = 0. \tag{5}$$

The significant approximation satisfying boundary condition (3b) has the form

$$U(\xi,y) = f(y)e^{-a\xi}. \tag{6}$$

(b) *The corner layer*

For $\alpha = \beta = 1$ the limiting equation is

$$\frac{\partial^2 W}{\partial \xi^2} + \frac{\partial^2 W}{\partial \eta^2} + a\,\frac{\partial W}{\partial \xi} - \frac{\partial W}{\partial \eta} = 0 \tag{7}$$

The boundary values for $W(\xi,\eta)$ are

$$W(\xi,0) = 0, \qquad W(0,\eta) = f(0). \tag{8}$$

Moreover the function $W(\xi,\eta)$ is required to satisfy the matching condition

$$W(\xi,\eta) = f(0)e^{-a\xi} \qquad \text{for } \eta \to \infty. \tag{9}$$

We obtain the following expression for $W(\xi,\eta)$:

$$W(\xi,\eta) = \frac{4f(0)}{\pi i}\, e^{-\frac{a\xi}{2}} \int_{-\infty}^{\infty} e^{-\frac{a\xi}{2}\sqrt{a^2+b^2+4\lambda^2} + \frac{\eta}{2}(b+2i\lambda)}\, \frac{\lambda}{4\lambda^2+b^2}\, d\lambda. \tag{10}$$

3. THE TRANSITION BOUNDARY LAYER

For $a = a_\gamma \varepsilon^\gamma$, $0 < \gamma < 1/2$ we observe the following modifications. The corner layer solution follows from (10) by taking $a = 0$. The ordinary boundary layer increases and will have a thickness of order $O(\varepsilon^{1-\gamma})$. Moreover a new layer arises:

(c) *The intermediate layer*

The function $X(\xi,\eta)$ has to satisfy the limiting equation

$$\frac{\partial^2 X}{\partial \xi^2} + a_\gamma \frac{\partial X}{\partial \xi} - b\,\frac{\partial X}{\partial \eta} = 0 \qquad (\alpha = 1-\gamma,\ \beta = 1-2\gamma). \tag{11}$$

The matching conditions are

$$\lim_{\varepsilon\to 0} W(\varepsilon^{\alpha-1}\xi, \varepsilon^{2(\alpha-1)}\eta) = \lim_{\varepsilon\to 0} X(\varepsilon^{\alpha-1+\gamma}\xi, \varepsilon^{2(\alpha-1+\gamma)}\eta) \quad \text{for } 1-\gamma < \alpha < 1 \tag{12}$$

and

$$\lim_{\eta\to\infty} X(\xi,\eta) = f(0)e^{-a_\gamma \xi}. \tag{13}$$

It appears that the solution of this problem already has some properties of the well-known parabolic boundary layer solution for characteristic boundaries:

$$X(\xi,\eta) = \frac{f(0)}{2}\{e^{-a_\gamma \xi}\,\mathrm{erfc}\Big(\frac{\xi}{2}\sqrt{\frac{b}{\eta}} - \frac{a_\gamma}{2}\sqrt{\frac{\eta}{b}}\Big) + \mathrm{erfc}\Big(\frac{\xi}{2}\sqrt{\frac{b}{\eta}} + \frac{a_\gamma}{2}\sqrt{\frac{\eta}{b}}\Big)\}. \tag{14}$$

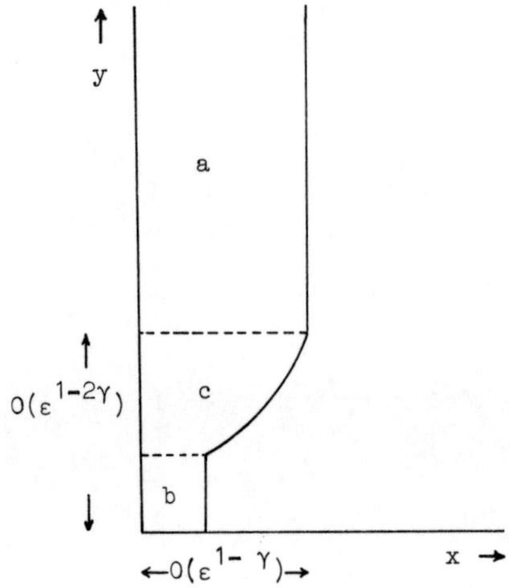

Fig. 2

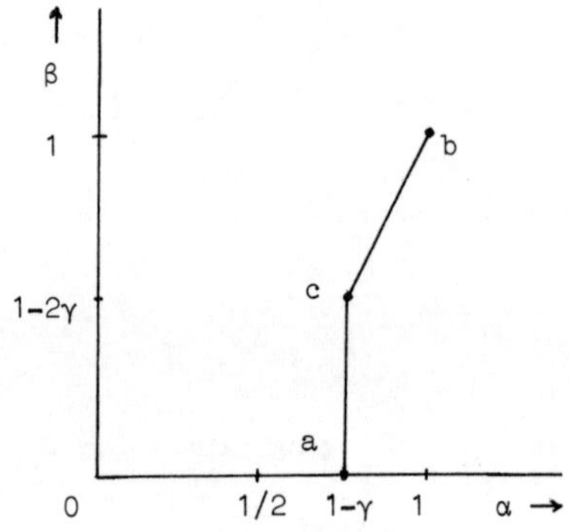

Fig. 3

4. THE PARABOLIC BOUNDARY LAYER

For $\gamma = 1/2$ the ordinary boundary layer vanishes and the intermediate layer transforms into

(d) *The parabolic boundary layer*

The parabolic boundary layer solution has to satisfy the limiting equation

$$\frac{\partial^2 Z}{\partial \xi^2} + a_{\frac{1}{2}} \frac{\partial Z}{\partial \xi} - b \frac{\partial Z}{\partial y} - cZ = 0, \qquad (\alpha=1/2,\ \beta=0) \tag{15}$$

and the boundary conditions

$$Z(\xi,0) = 0, \qquad Z(0,y) = f(y), \tag{16}$$

which yields

$$Z(\xi,y) = \frac{\xi\sqrt{b}\, e^{-\frac{a_{\frac{1}{2}}}{2}\xi}}{2\sqrt{\pi}} \int_0^y \frac{f(p)\, e^{\frac{-\xi^2 b}{4(y-p)} - (\frac{1}{4}a_{\frac{1}{2}}^2+c)(y-p)}}{(y-p)^{3/2}}\, dp. \tag{17}$$

Since $Z(\xi,y)$ remains regular for $a_{\frac{1}{2}} \to 0$, we may conclude that the parabolic boundary layer solution is represented by (17) in all cases where the boundary coincides with the subcharacteristic of L_1 with an accuracy of $O(\sqrt{\varepsilon})$.

REFERENCES

Eckhaus, W., 1968, On the structure of certain asymptotic boundary layers, Report Math. Inst. Delft Univ. Techn..

Eckhaus, W., and E.M. de Jager, 1966, Asymptotic solutions of singular pertubation problems for linear differential equations, Arch. Rat. Mech.

Anal. 23, p. 26.
Grasman, J., 1971, On the birth of boundary layers, Math. Cent. Tract no. 36, Mathematisch Centrum Amsterdam.
Grasman, J., 1974, An elliptic singular perturbation problem with almost characteristic boundaries, to appear in J. Math. Anal. Appl..
Visik, M.I., and L.A. Lyusternik, 1962, Regular degeneration and boundary layer for linear differential equations with small parameter, Amer. Math. Soc. Trans. Ser. 2, 20, p. 239.

SINGULAR PERTURBATION PROBLEMS FOR NON-LINEAR ELLIPTIC SECOND ORDER EQUATIONS.

A. VAN HARTEN
Mathematical Institute
University of Utrecht,
The Netherlands.

Abstract.
We study elliptic problems of the form $F(x_i,u,u_i,\varepsilon\cdot u_{ij})=f$ with Dirichlet boundary conditions. A formal approximation Z is supposed to be given such that the equation and boundary conditions are satisfied up to the order $\chi(\varepsilon) = o(1)$, in some norm. It will be shown that under certain conditions one can prove the difference between the exact solution u and the formal approximation Z to be small: $O(\overline{\chi}(\varepsilon)) = o(1)$, where $\overline{\chi}$ of course will depend on χ. Such proofs will be based on the maximum principle for non-linear elliptic equations or on a contraction principle in a suitable Banach space.

Contents.
1. Introduction.
2. Some norm estimates for linear elliptic singular perturbation problems.
3. Applications of the maximum principle for non-linear elliptic equations.
4. Application of the contraction principle in a suitable Banach space.
5. List of references.

1. INTRODUCTION.

1.1. The most general form of a second order nonlinear equation in n variables containing a small parameter $\varepsilon>0$ can be given as:

$$F(x_i,u,u_i,u_{ij}|\varepsilon) = f \tag{1.1}$$

here $x_1,\ldots,x_n$ are the independent variables and u is the dependent variable.
The derivatives $\frac{\partial u}{\partial x_i}$, $\frac{\partial^2 u}{\partial x_i \partial x_j}$ are denoted by u_i, u_{ij}. The functional operator F and the inhomogeneous term f are supposed to be analytical for all the significant values of their variables and F is taken symmetrical with respect to the u_{ij}'s.
D will be an open domain $\subset \mathbb{R}^n$ with a closure $\overline{D}$ and a sufficiently smooth boundary $S = \overline{D}\backslash D$. The operator F will be called uniformly elliptic in D if for all $\vec{x} \in \overline{D}$ and all values of u, u_i, u_{ij} and ξ_i

$$\sum_i \sum_j \frac{\partial F}{\partial u_{ij}} \cdot \xi_i \cdot \xi_j \geqslant A \cdot \sum_i \xi_i^2 \tag{1.2}$$

A is some constant > 0, in general dependent of ε.
In this paper we shall consider uniformly elliptic problems on a bounded domain D with boundary conditions of the Dirichlet type:

(1.3) $u = \phi$ on S

with ϕ sufficiently differentiable and independent of ε. We shall call 1.1. a singular perturbation problem if the operator has the form:

(1.4) $F = F(x_i, u, u_i, \varepsilon\, u_{ij})$

We shall pay special attention to some examples from the class of semi-linear singular perturbation problems:

(1.5) $\varepsilon \cdot L_2 u + g(x_i, u, u_i) = f$
$u = \phi$ on S

where L_2 is a linear, uniformly elliptic operator.
In what follows we shall frequently use the following notations:
For the cartesian vectornorm: $\|\vec{x}\| = \{\sum_i x_i^2\}^{\frac{1}{2}}$

Let $h(\vec{x})$ be sufficiently differentiable on $\bar{D}$. Then:

(1.6) $[h]_m = \max_{|k|=m} \sup_{\vec{x}\in\bar{D}} |h^{(k)}(x)|$; $m = 0,1,2,\ldots$

(1.7) $[h]_{m+\alpha} = \max_{|k|=m} \sup_{\vec{x},\vec{y}\in\bar{D}} \dfrac{|h^{(k)}(\vec{x}) - h^{(k)}(\vec{y})|}{\|\vec{x} - \vec{y}\|^{\alpha}}$; $0<\alpha<1$.

In 1.6-7 k represents a multi-index:

$k = (k_1,\ldots,k_n) \in \mathbb{N}^n$; $|k| = \sum_{i=1}^{n} k_i$.

We introduce the Höldernorms:

(1.8) $|h|_m = \sum_{j=0}^{m} [h]_j$

(1.9) $|h|_{m+\alpha} = |h|_m + [h]_{m+\alpha}$.

Remark that the maximumnorm is denoted by $|\ |_0$.
The analogous norms for functions defined on the boundary S will be denoted by an upperprime. So:

(1.10) $|g|_0' = \sup_{\vec{x}\in S} |g(\vec{x})|$

etc.

1.2. Analogously to linear elliptic singular perturbation problems it will be possible for a number of non-linear cases to construct a sequence $Z_m(\vec{x},\varepsilon)$ with the following properties

(1.11a) $F(Z_m) - f = r_m$

(1.11b) $\phi - Z_m = s_m$

where in some norm $|\ |$ on D and $|\ |'$ on S:

(1.12a) $|r_m| = O(\delta_m(\varepsilon))$

(1.12b) $|s_m|' = O(\delta_m'(\varepsilon))$

The orders δ_m, δ'_m will improve with increasing m. For the construction of such sequences Z_m in linear cases we refer to the survey in [1] or [2].

We whall consider here the following question: given a sequence Z_m with the properties specified above, how to prove that the Z_m's indeed represent approximations of a solution u in some suitable norm $\| \ \|$ with some accuracy $\tilde{\delta}_m(\varepsilon)$? So:

(1.13) $\quad \|u - Z_m\| = O(\tilde{\delta}_m(\varepsilon))$??

One way to give such proofs is to apply a maximumprinciple for non-linear elliptic equations. This will be demonstrated in section 3. The method was originally used for linear cases by Eckhaus and de Jager, [3]. If this method fails, one sometimes can prove a result like 1.13 using a contraction principle in a suitable Banach space. This will be shown in section 4.

As a preliminary we shall treat first some theorems concerning linear elliptic equations in section 2. The reason is that we shall need them further on, and that they represent examples valid in the linear case of what we want to prove for non-linear equations.

2. SOME NORM-ESTIMATES FOR LINEAR ELLIPTIC SINGULAR PERTURBATION PROBLEMS.

2.1. The linear problem is

(2.1a) $\quad L_\varepsilon \cdot u = \{\varepsilon \cdot L_2 + L_1\} \cdot u = h$

(2.1b) $\quad u = \phi$ on S

L_2 is uniformly elliptic in $\overline{D}$ and independent of ε and L_1 is a first order operator:

(2.2) $\quad L_1 = \sum_{i=1}^{n} \beta_i(\vec{x},\varepsilon) \frac{\partial}{\partial x_i} + \gamma(\vec{x},\varepsilon)$

In order to be able to use the maximumprinciple we suppose in this section that L_1 satisfies the following condition: there exist constants $\vec{\theta} \in \mathbb{R}^n$ and $\delta > 0$ independent of ε such that:

(2.3) $\quad \vec{\theta} \cdot \vec{\beta} + \gamma \leqslant - \delta$

where $\vec{\theta} \cdot \vec{\beta}$ is an innerproduct notation.

Now the following result on barrierfunctions is valid, as can be seen by transforming the problem above into one where the maximumprinciple of ref. [4] can be applied:

<u>Theorem</u> 2.i.

Given z and Z such that:

(a): $L_\varepsilon \cdot Z \leqslant h \leqslant L_\varepsilon \cdot z$ in $\overline{D}$

(b): $z \leqslant \phi \leqslant Z$ on S

Then, for ε small enough, $0 < \varepsilon < \varepsilon_0$:

$$z \leqslant u \leqslant Z$$

As a consequence of theorem 2.i one finds the following norm estimates:

Theorem 2.ii.

There exists a constant $C > 0$ independent of ε such that for ε small enough, $0 < \varepsilon < \varepsilon_0$: $|u|_0 \leqslant C \cdot \{|h|_0 + |\phi|_0'\}$

If in the linear case a sequence Z_m is given with the properties (1.11), (1.12) where $|\ | = |\ |_0$ and $|\ |' = |\ |_0'$ then one finds with theorem 2.ii

(2.4) $\quad |u-Z_m|_0 \leqslant C \cdot \{|r_m|_0 + |s_m|_0'\} = O(\max(\delta_m, \delta_m'))$

This then answers question (1.13).

2.2. The linear problem of the equations 2.1a-b can be reduced to a problem with homogeneous boundary conditions by the following wellknown transformation to the new dependent variable:

(2.5a) $\quad w = u - W$, with:

(2.5b) $\quad W = \phi$ on S

so:

(2.6a) $\quad L_\varepsilon \cdot w = f = h - L_\varepsilon \cdot W$

(2.6b) $\quad w = 0$ on S

The boundary S and the boundary values ε are now supposed to be ∞-differentiable. Then we can choose W ∞-differentiable and such that for $0 \leqslant \sigma \leqslant 2 + \alpha$ with $0 < \alpha < 1$:

(2.7) $\quad |W|_\sigma \leqslant C_0 \cdot |\phi|_\sigma$

C_0 is a constant $\geqslant 1$ independent of ε.

In this subsection we shall derive some estimates for the norms $|w_\sigma|$; $0 < \sigma \leqslant 2+\alpha$ where w satisfies 2.8a-b in terms of f, which imply estimates of u in terms of h and ϕ of course. These estimates are based on the following theorem of Agmon, Douglas and Nirenberg, ref. [5]:

Theorem 2.iii.

The solution v of the problem:

(2.8) $\quad M \cdot v = g$, $\quad v = 0$ on S

where M is uniformly elliptic in $\overline{D}$, satisfies the inequality:

(2.9) $\quad |v|_{2+\alpha} \leqslant C \cdot \{|g|_\alpha + |v|_0\}$

For the case $M = L_\varepsilon$ one is able to determine the dependence on ε of the constant C in 2.9 by a method similar to the one applied by Besjes, ref. [6]. If theorem 2.ii is used to eliminate the term $|w|_0$ and some inequalities of calculus are used in order to estimate $|w|_\gamma$ in terms of $|w|_{2+\alpha}$ and $|w|_0$ (,ref. [7]) then the following theorems can be derived:

Theorem 2.iv.

If there exists a number $\alpha > 0$ such that for $0 \leqslant \sigma \leqslant \alpha$:

a. $|\beta_i|_\sigma \leqslant B \cdot \varepsilon^{-\sigma}$, $i = 1,\ldots,n$

b. $|\gamma|_\sigma \leqslant C \cdot \varepsilon^{-1-\sigma}$

with constants $B,C > 0$ independent of ε, then we have for $0 \leqslant \gamma \leqslant 2+\alpha$ the following estimate:

$$|w|_\gamma \leqslant D \cdot \varepsilon^{-\gamma} \cdot |f|_\alpha$$

with a constant $D > 0$ independent of ε.

and:

Theorem 2.v.

If there exists a number $\alpha > 0$ such that for $0 \leqslant \sigma \leqslant \alpha$:

a. $\beta_i = 0$, $i = 1,\ldots,n$

b. $|\gamma|_\sigma \leqslant C \cdot \varepsilon^{-\sigma/2}$

with a constant $C > 0$, independent of ε, then we have for $0 \leqslant \gamma \leqslant 2+\alpha$ the following estimate:

$$|w|_\gamma \leqslant D \cdot \varepsilon^{-\gamma/2} \cdot |f|_\alpha$$

with a constant $D > 0$, independent of ε.

Here β_i, $i = 1,\ldots,n$ and γ are the coefficients of L_1 as specified in 2.2. We remark that the facts:

(2.10a) $|w|_\gamma = O(\varepsilon^{-\gamma})$ if $\beta_i \neq 0$ for some i

(2.10b) $|w|_\gamma = O(\varepsilon^{-\gamma/2})$ if $\beta_i = 0$ for all i

if β_i, γ and f are independent of ε, are in agreement with the boundary layer structure found in constructions (ref. [1], [2]).

3. APPLICATIONS OF THE MAXIMUM PRINCIPLE FOR NON-LINEAR ELLIPTIC EQUATIONS.

3.1. Suppose a function Z is given satisfying:

(3.1a) $F(Z) = f + r$

(3.1b) $Z = \phi + s$ on S

where r and s are perturbations, with:

(3.2) $|r|_0 + |s|'_0 = O(\chi(\varepsilon)) = o(1)$.

Our aim is to show that we can estimate $|u - Z|_0$ under certain conditions by application of the technique of barrierfunctions. Here u is the exact solution of 1.1, 1.3, which is assumed to exist. This exact solution will automatically be unique as a consequence of the theorems on barrierfunctions that we shall give hereafter.

Let z_l, z_u be given such that:

(3.3a) $F(z_u) \leqslant f \leqslant F(z_l)$ in $\overline{D}$

(3.3b) $z_l \leqslant \phi \leqslant z_u$ on S

We define the class of functions:

(3.4) $H = \{v | v = u+\lambda(u-z_l) \text{ or } v = u+\lambda(u-z_u);\ 0\leqslant\lambda\leqslant 1\}$.

Next we formulate:

<u>Theorem</u> 3.i.

If: $F_u(v) \leqslant 0$, $\forall\ v \in H$ then: $z_l \leqslant u \leqslant z_u$ in $\overline{D}$.

<u>Proof</u>. ref [4].

In the case that F is semi-linear (see 1.5) a somewhat different version of theorem 3.i will also be valid:

<u>Theorem</u> 3.ii.

If there exist constants $\theta_1,\ldots,\theta_n$ and $\delta > 0$ independent of ε such that:

$$\sum_{i=1}^{n} \theta_i \cdot F_{u_i}(v) \leqslant -\delta \text{ for } v \in H$$

then: $z_l \leqslant u \leqslant z_u$ in $\overline{D}$.

<u>Proof</u>.

The proof can be given analogous to ref. [4] by using theorem 2.i.

We now apply theorem 3.i to a pair of barrier functions:

(3.5a) $z_l = Z + \chi(\varepsilon) \cdot \omega$

(3.5b) $z_u = Z + \chi(\varepsilon) \cdot \Omega$

with ω and Ω suitably chosen to prove the following theorem.

<u>Theorem</u> 3.iii.

If (a): $F_u \leqslant 0$.

(b): $F_u(Z) \leqslant -\delta < 0$, with δ independent of ε.

(c): $\chi(\varepsilon)$ is small enough to satisfy:

$$\left|\int_0^{\chi} (\chi-\xi) \cdot F_{uu}(Z+t\xi) \cdot d\xi\right|_0 = o(\chi)$$

with t a fixed constant independent of ε.

then: $|u-Z|_0 = O(\chi)$.

<u>Proof</u>:

In 3.5a-b we choose ω and Ω constant, such that 3.3a-b are satisfied. In order to verify 3.3a one applies a Taylorexpansion of F with respect to $\chi(\varepsilon)$.
We will investigate more explicitely under what conditions the requirement c of this theorem will be fulfilled.
Remark that:

(3.6) $$\left|\int_0^{\chi}(\chi-\xi)\cdot F_{uu}(Z+t\cdot\xi)\cdot d\xi\right|_0 \leqslant \tfrac{1}{2}\cdot\chi^2\cdot \max_{|\xi|\leqslant t\cdot\chi} |F_{uu}(Z+\xi)|_0$$

Suppose there exists a constant $\eta > 0$, independent of ε, such that:

$$(3.7) \qquad \max_{|\xi|\leqslant\eta} |F_{uu}(Z+\xi)|_0 \leqslant \sigma^{-1}(\varepsilon).$$

It is easily verified that 3.iii-c is fulfilled, if:

$$(3.8) \qquad \chi(\varepsilon) = o(\sigma(\varepsilon))$$

So in the situation that $\sigma(\varepsilon)$ does not depend on ε requirement 3.iii-c is fulfilled automatically because of the supposition of 3.2: $\chi = o(1)$.
Further we remark that in condition (a) we require $F_u \leqslant 0$ everywhere. Of course everywhere can be replaced here by: for functions of the class H of 3.4 with z_u, z_1 as indicated in the proof. Next we shall give theorem 3.iv which generalizes theorem 3.iii and can be proved analogously.

Theorem 3.iv.

If: (a) $F_u \leqslant 0$

(b) $\sum_{i=1}^{n} \Theta_i \cdot F_{u_i}(Z) + F_u(Z) \leqslant -\delta < 0$
with constants $\Theta_1,\ldots,\Theta_n,\delta$ independent of ε.

(c) $\chi(\varepsilon)$ is small enough to satisfy

$$\left|\int_0^{\chi}(\chi-\xi) \cdot \frac{\partial^2}{\partial\xi^2} F(Z + \xi \cdot 1) \cdot d\xi\right|_0 = o(\chi)$$

for a linear function $1 = t_1 \cdot (\vec{\Theta} \cdot \vec{x} + t_2)$
with fixed constants t_1, t_2 independent of ε,

then: $|u - Z|_0 = O(\chi)$.

In the same way as before we find that the requirement 3.iv-c is fulfilled, if for each linear function $1 = a \cdot (\vec{\Theta}\cdot\vec{x} + b)$ with $|a| \leqslant A$; $|b| \leqslant B$

$$(3.9) \qquad |F_{uu}(Z+1)|_0 + \sum_{i=1}^{n} |F_{uu_i}(Z+1)|_0 + \sum_i \sum_j |F_{u_iu_j}(Z+1)|_0 \leqslant \sigma^{-1}(\varepsilon)$$

and

$$(3.10) \qquad \chi(\varepsilon) = o(\sigma(\varepsilon))$$

A, B are constants > 0 independent of ε. The same remark on condition (a) as before applies here also. The last theorem of the same sort as 3.iii-iv states:

Theorem 3.v.

If: (a) F is semilinear

(b) $\sum_{i=1}^{n} \Theta_i \cdot F_{u_i} + F_u \leqslant -\delta < 0$
with constants $\Theta_1,\ldots,\Theta_n$ and δ independent of ε

(c) $\chi(\varepsilon)$ is small enough to satisfy

$$|\int_0^{\chi}(\chi-\xi)\cdot\frac{\partial^2}{\partial\xi^2}F(Z+\xi\cdot e)\cdot d\xi|_0 = o(\chi)$$

for $e = c \cdot \exp(\vec{\Theta}\cdot\vec{x})$

with a fixed constant c independent of ε.

then: $|u - Z|_0 = O(\chi)$

Proof:

If Ω, ω are chosen of the form: $c \cdot \exp(\vec{\Theta}\cdot\vec{x})$ with a suitable constant c then 3.3a-b are fulfilled. Theorem 3.ii now completes the proof.

It will be clear that condition c is fulfilled if 3.9, 3.10 are valid with 1 replaced by $e = c \cdot \exp(\vec{\Theta}\cdot\vec{x})$ with $|c| \leq C$. Here C represents again a constant > 0 independent of ε. The remark made for 3.iii-a applies here to condition b. Finally we mention without working out that further generalizations of the theorems 3.iii-v are possible for example to cases where δ depends on ε.

3.2. We investigate now under what conditions the theorems of the preceding subsection can be applied to some explicit examples.
We work in these examples in $\mathbb{R}^2$ where the coordinates are x and y.

Example A.

(3.11) $$\varepsilon \cdot L_2 u - k \cdot u_y - g(u) = f$$
$$u = \phi \text{ on } S$$

with a constant $k \in \mathbb{R}$.

We assume the constructed formal approximation Z to be bounded by bounds independent of ε, so

(3.12) $M_1 \leq Z \leq M_2$

and furthermore g is ∞-differentiable.

We distinguish between the cases: $k = 0$; $k \neq 0$

(a) $k = 0$.

Theorem 3.iii can be applied here, if

(3.13) $g' \geq 0$ and $g'(Z) \geq \delta > 0$

Condition 3.iii-c will be satisfied automatically according to 3.7, 3.8.
For example 3.13 is fulfilled by the choice:

$g = u + u^3$; $g = \exp(u)$, etc.

Of course it is not necessary in 3.13 to suppose that

$g' \geq 0$ everywhere as the following example shows. Suppose:

(3.14) $$\begin{cases} g = u + u^2 \\ \phi \geq -\frac{1}{2} + \alpha \text{ and } f \leq -\frac{1}{4} + \alpha^2 \text{ with } \alpha > 0 \end{cases}$$

then: $-\frac{1}{2} + \alpha \leq u \leq M$ with M sufficiently large, because of theorem 3.i with constant barrierfunctions.

Surely also: $-\frac{1}{2} + \frac{1}{2}\alpha \leqslant Z \leqslant M + 1$.
For the class H of 3.4 the requirement $g' > 0$ will be fulfilled. So in this example theorem 3.iii can be applied under suitable conditions for f and ϕ such as in 3.14. This situation arises more frequently, for example:

(3.15) $$\begin{cases} g = \sin(u) \\ |\phi| \leqslant \alpha \text{ and } |f| \leqslant \sin(\alpha) \text{ with } 0 < \alpha < \frac{\Pi}{2} \end{cases}$$

etc.

(b) <u>$k \neq 0$</u>.

if:

(3.16) $g' \geqslant -C$ with some constant $C \in \mathbb{R}$, independent of ε.

then theorem 3.v can be applied directly, because with $\Theta_1 = 0$, $\Theta_2 = (|C|+1)/k$, 3.v-b is fulfilled and 3.v-c is fulfilled automatically.

Again we remark that even in this very simple example the existence of u still is assumed. This assumption will be justified in section 4.

Example A was one out of the class of semi-linear equations. Theorem 3.i can be applied at once to the non semi-linear case:

<u>Example</u> B.

(3.17) $$\begin{cases} \varepsilon \, , \, \{(1+u_y^2)u_{xx} - 2u_x u_y u_{xy} + (1+u_x^2)\, u_{yy}\} - u = f \\ u = \phi \text{ on } S \end{cases}$$

which is an equation related to the minimal surface equation.

In the following example none of the theorems of 3.1 can be used.

<u>Example</u> C.

(3.18) $$\varepsilon \cdot L_2 u - P(u) \cdot u_y = f$$
$$u = \phi \text{ on } S$$

We investigate:

$$\Theta \cdot F_{u_y} + F_u = -\Theta \cdot P(u) - P'(u) \cdot u_y$$

Even if we assume for example: $P(u) \geqslant \delta > 0$ nothing can be said about this expression without a priori information about the sign of u_y. This information is not so easy obtainable.

As we will see, the equation 3.18 can be handled by the method of the following section.

4. APPLICATION OF THE CONTRACTION PRINCIPLE IN A SUITABLE BANACH SPACE.

4.1. The theory as given in this subsection will be valid for more general problems of second order. We can formulate our problem as:

(4.1a) $F_\varepsilon(u) = f$

with initial- and \ or boundary conditions:

(4.1b) $\vec{K}(u) = \vec{\phi}$

where $\vec{K}$ is supposed to be linear.

Again a formal approximation Z is supposed to be given, which is sufficiently differentiable and which satisfies:

(4.2) $\begin{cases} F_\varepsilon(Z) = f + r \\ \vec{K}(Z) = \quad + s \end{cases}$

and

(4.3) $|r| + |\vec{s}|' \leqslant \chi(\varepsilon) = o(1)$

$|\ |$ and $|\ |'$ are suitable (semi-) norms. We define the linearized operator in Z by:

(4.4) $L_\varepsilon = \sum_{i,j} \frac{\partial F_\varepsilon}{\partial u_{ij}}(Z) \cdot \frac{\partial^2}{\partial x_i \partial x_j} + \sum_i \frac{\partial F_\varepsilon}{\partial u_i}(Z) \cdot \frac{\partial}{\partial x_i} + \frac{\partial F_\varepsilon}{\partial u}(Z)$

and the remainder $\Psi_\varepsilon(R)$ by:

(4.5) $\Psi_\varepsilon(R) = F_\varepsilon(Z+R) - F_\varepsilon(Z) - L_\varepsilon \cdot R$

$\|\ \|$ will be <u>a</u> norm defined at least for ∞-differentiable function on $\bar{D}$. W is a Banach space with respect to the norm $\|\ \|$. Hereafter we will specify three conditions which guarantee the existence of a solution of 4.1 within some neighbourhood of radius $\omega(\varepsilon) = o(1)$ of $Z \in W$ in the sense of the norm $\|\ \|$, so:

(4.6) $\|u - Z\| \leqslant \omega(\varepsilon)$

The first condition concerns the solutions of the linearized problem in Z.

<u>Condition</u> a:

The problem

(4.7) $\begin{cases} L_\varepsilon \cdot R = r \\ \vec{K}(R) = \vec{s} \end{cases}$

has a unique solution $R = L_\varepsilon^{-1} \cdot r \in W$ which satisfies

(4.8) $\|L_\varepsilon^{-1} \cdot r\| \leqslant l(\varepsilon)^{-1} \cdot \{|r| + |s|'\}$

Secondly the remainder term Ψ_ε has to be "quadratic" in the following sense:

<u>Condition</u> b:

$\exists m(\varepsilon)$ such that for all ρ, $0 < \rho \leqslant \rho_0$ and for all R_1, R_2 with $\|R_1\| < \rho$ and $\|R_2\| < \rho$:

(4.9) $|\Psi_\varepsilon(R_1) - \Psi_\varepsilon(R_2)| \leqslant m(\varepsilon)^{-1} \cdot \rho \cdot \|R_1 - R_2\|$

Finally we require $\chi(\varepsilon)$ to be small enough.
Condition c:
for ε sufficiently small:

(4.10) $\chi(\varepsilon) < \frac{1}{4} \cdot l(\varepsilon)^2 \cdot m(\varepsilon)$

The following result now can be proved by using the contraction principle in the Banach space W.
Theorem 4.i.
If a, b and c are satisfied then there exists one solution u of 4.1 with:

(4.11) $\|u - Z\| \leq \omega(\varepsilon) = 2 \cdot \chi(\varepsilon) \cdot l(\varepsilon)^{-1}$.

Proof:
We rewrite the problem for u = Z + R as:

(4.12) $\begin{cases} L_\varepsilon \cdot R = -r - \Psi_\varepsilon(R) \\ \vec{K}(R) = \vec{s} \end{cases}$

or as a search for a fixed point:

(4.13) $R = T_\varepsilon(R) = L_\varepsilon^{-1}\{-r - \Psi_\varepsilon(R)\}$

One easily shows that:
(i) T_ε is a strict contraction on the sphere $B(\omega(\varepsilon)) = \{\|R\| \leq \omega(\varepsilon)\}$
(ii) $T_\varepsilon(B(\omega(\varepsilon))) \subset B(\omega(\varepsilon))$.

It is a wellknown fact that a strictly contractive map T_ε from a convex subset $B(\omega(\varepsilon))$ of a Banach space W into itself possesses exactly one fixed point R_0 within $B(\omega(\varepsilon))$. So this R_0 is the unique solution of 4.12 within $B(\omega(\varepsilon))$ and with $u = Z + R_0$ the theorem is proved.

We remark that the concept of solution is used in this theorem in a generalized sense. After all, the elements of W are not necessarily 2 times continuously differentiable. The result $\omega(\varepsilon) = O(\chi(\varepsilon) \cdot l(\varepsilon)^{-1})$ is of course the sharpest one that is possible because this is already found for the linearized problem. It will be found if the remainder term Ψ_ε caused by the non-linearity in 4.12 is neglectable if compared with the perturbations r, $\vec{s}$ of the order $\chi(\varepsilon)$, so if:

(4.14) $m(\varepsilon) \cdot \omega(\varepsilon)^2 = o(\chi(\varepsilon)) \Rightarrow$
$m \cdot l^{-2} \cdot \chi^2 = o(\chi) \Rightarrow$
$\chi = o(l^2 \cdot m)$

This clarifies condition c.
Next we consider some applications of the theorem 4.i to elliptic singular perturbation problems. We remark that in a somewhat different way the technique of using a contraction theorem in this field has been applied earlier by Berger & Fraenkel, ref. [8], [9].

4.2. (i) In a first application we consider again the example A of section 3.2 (see 3.11-12). In this case one finds:

(4.15) $L_\varepsilon \cdot R = \varepsilon \cdot L_2 R - k \cdot \frac{\partial R}{\partial y} - g'(Z) \cdot R$

(4.16) $\Psi_\varepsilon(R) = -g(Z+R) + g(Z) + g'(Z)\cdot R$

It will not be surprising that we work with the norms:

(4.17) $|\ | = \|\ \| = |\ |_0$ and $|\ |' = |\ |_0'$

So:

(4.18) $W = C(\overline{D})$

We shall verify that the conditions a, b and c of theorem 4.i are satisfied if:

(4.19) (a) $k = 0$ and $g'(Z) \geq \delta > 0$
or
(b) $k \neq 0$

Condition a:

for the problem $\begin{cases} L_\varepsilon \cdot R = r \\ R = s \text{ on } S \end{cases}$

2.3 is satisfied because of 4.19. So we obtain by theorem 2.ii:

$$|R|_0 \leq l^{-1}\cdot\{|r|_0 + |s|_0'\}$$

with a constant $l > 0$ independent of ε.

Condition b:

We derive by repeated application of the mean-value theorem:

$$\begin{aligned}\Psi_\varepsilon(R_1)-\Psi_\varepsilon(R_2) &= g(Z+R_2)-g(Z+R_1)+g'(Z)\cdot\{R_1-R_2\}\\ &= \{-g'(Z+R_1+\lambda(R_2-R_1))+g'(Z)\}\cdot\{R_1-R_2\}\\ &= g''(Z+\lambda'(R_1+\lambda(R_2-R_1)))\cdot\{R_1+\lambda(R_2-R_1)\}\\ &\qquad \{R_1-R_2\}\end{aligned}$$

with $0 \leq \lambda,\ \lambda' \leq 1$.

So, if $|R_1|_0$ and $|R_2|_0 \leq \rho$, then indeed:

$$|\Psi_\varepsilon(R_1) - \Psi_\varepsilon(R_2)|_0 \leq m^{-1}\cdot\rho\cdot|R_1 - R_2|_0$$

with $m > 0$ constant independent of ε.

Condition c:

$\chi(\varepsilon) \leq \frac{1}{4}\cdot l^2\cdot m$ for ε sufficiently small is fulfilled automatically because of: $\chi(\varepsilon) = o(1)$.

If we compare 4.19 with 3.13 and 3.16 respectively then we observe that theorem 4.i can be applied in this example under weaker conditions then the ones required to apply the theorems based on the maximumprinciple, while in both cases we prove:

(4.20) $|u - Z|_0 = O(\chi(\varepsilon))$.

Another advantage of theorem 4.i is that the existence of the solution u is proved, and needs not to be assumed as in section 3.2.

(ii) Secondly we consider again the example C of section 3.2 (see 3.18). We suppose S as well as ϕ ∞-differentiable.

For L_ε, Ψ_ε one finds:

(4.21) $L_\varepsilon\cdot R = \varepsilon\cdot L_2R - P(Z)\cdot R_y - P'(Z)\cdot Z_y\cdot R$

(4.22) $\Psi_\varepsilon(R = -P(Z+R)\cdot\{Z_y+R_y\}+P(Z)\cdot\{Z_y+R_y\}+P'(Z)\cdot Z_y\cdot R$

As a special case we consider:

(4.23) $\begin{cases} \varepsilon \cdot \Delta u - uu_y = 0 \\ u = \phi \text{ on the unit circle.} \end{cases}$

For this special case 4.21, 4.22 reduce to:

(4.24) $L_\varepsilon \cdot R = \varepsilon \cdot \Delta R - Z \cdot R_y - Z_y \cdot R$

(4.25) $\Psi_\varepsilon(R) = -R \cdot R_y$

We assume the formal approximation Z to be ∞-differentiable and:

(4.26) a. $|Z|_\beta \leqslant M \cdot \varepsilon^{-\beta}$

b. $Z_y \geqslant -M$

with $0 \leqslant \beta \leqslant 2$ and $M > 0$, independent of ε.
Further we suppose:

(4.27) $P(Z) \geqslant \delta > 0.$

with δ independent of ε.

In general 4.26, 4.27 can only be satisfied for suitable boundary conditions ϕ and inhomogeneous term f. In the special case 4.23 one can show by performing the construction of Z explicitly that 4.26, 4.27 are satisfied, if for $0 < x < 1$:

(4.28) $0 < \delta \leqslant \phi(x, -\sqrt{1-x^2}) < \phi(x, \sqrt{1-x^2}).$

Of course we can take by addition of a suitable ∞-differentiable function W as in 2.7-9, Z such that:

(4.29) $Z = \phi$ on S

So, in 4.2: $s = 0$. We take as norms:

(4.30) $\| \ \| = | \ |_{1+\alpha}; \ | \ | = | \ |_\alpha$

and as our Banachspace:

(4.31) $W = \{f | f \in C^{1+\alpha}(\bar{D}) \text{ and } f = 0 \text{ on } S\}.$

α is a small number > 0.

We shall investigate again the conditions a, b and c.

Condition a.

for the problem $\begin{cases} L_\varepsilon \cdot R = r \\ R = 0 \text{ on } S \end{cases}$

2.3 and 2.iv.a-b are satisfied because of 4.26 -27. So we obtain by theorem 2.iv:

$$|R|_{1+\alpha} \leqslant C \cdot \varepsilon^{-1-\alpha} \cdot |r|_\alpha$$

So condition a is satisfied with:

$$l(\varepsilon) = C \cdot \varepsilon^{1+\alpha}.$$

Here and furtheron C is a suitable constant > 0 independent of ε.

Condition b.

In the special case 4.25 one obtains:

$$\begin{aligned} \Psi_\varepsilon(R_1)-\Psi_\varepsilon(R_2) &= -R_1 \cdot R_{1,y}+R_2 \cdot R_{2,y} \\ &= -R_1 \cdot R_{1,y}+R_1 \cdot R_{2,y}-R_1 \cdot R_{2,y}+R_2 \cdot R_{2,y} \\ &= R_1 \cdot (R_{2,y}-R_{1,y})+R_{2,y} \cdot (R_2-R_1) \end{aligned}$$

If $|R_1|_{1+\alpha} \leqslant \rho$ and $|R_2|_{1+\alpha} \leqslant \rho$ then:

$$|\Psi_\varepsilon(R_1)-\Psi_\varepsilon(R_2)|_\alpha \leqslant 2 \cdot \rho \cdot |R_1-R_2|_{1+\alpha}.$$

So in this special case condition b is satisfied with $m(\varepsilon) = \frac{1}{2}$.
In the general case one can derive by a lengthy calculation, using a trick like the one above, the mean value theorem as in the verification of condition b in section 4.2.i and the estimates of 4.26, that:

$$|\Psi_\varepsilon(R_1)-\Psi_\varepsilon(R_2)|_\alpha \leq C \cdot \varepsilon^{-1-\alpha} \cdot \rho \cdot |R_1-R_2|_{1+\alpha}$$

So, then condition b is satisfied with: $m(\varepsilon) = C \cdot \varepsilon^{1+\alpha}$.
Condition c.
for ε sufficiently small we must have:

(4.32) $\chi(\varepsilon) \leq C \cdot \varepsilon^{3+3\alpha}$

in the general case and in the special case 4.23:

(4.33) $\chi(\varepsilon) \leq C \cdot \varepsilon^{2+2\alpha}$.

So in practive the construction of such a function Z will be quite laborious! The result of the application of theorem 4.i now is: if 4.26, 27, 29 and 4.32 (or 4.33 in the special case of 4.23) then there is one solution u with:

(4.34) $|u - Z|_{1+\alpha} \leq C \cdot \chi(\varepsilon) \cdot \varepsilon^{-1-\alpha}$.

Once this result is obtained one can derive an estimate for $|u-Z|_0$ in the following way: for R = u-Z we have (as in 4.11):

$$\begin{cases} L_\varepsilon \cdot R = - r - \Psi_\varepsilon(R) \\ R = 0 \text{ on } S \end{cases}$$

Now:

$$|r + \Psi_\varepsilon(R)|_0 \leq |r + \Psi_\varepsilon(R)|_\alpha \leq C \cdot \chi(\varepsilon)$$

So by theorem 2.ii:

(4.35) $|u - Z|_0 \leq C \cdot \chi(\varepsilon)$.

We remind however that this result is valid under the same conditions as 4.34 and $\chi(\varepsilon)$ is an estimate of the remainder term in the norm $|\ |_\alpha$!

Finally the author wants to express his gratitude indebted to Professor W. Eckhaus for his most helpful and inspiring discussions about this work.

[1]. Eckhaus, W. : "Matched asymptotic expansions and singular perturbations", 1973, North-Holland/American Elsevier.

[2]. Eckhaus, W. : "Boundary layers in linear elliptic singular perturbation problems", 1972, Siam Review, vol. 14, p. 225 - 270.

[3]. Eckhaus, W.; De Jager, E.M. : "Asymptotic solutions of singular perturbation problems for linear differential equations", 1966, Arch. for rat. mech. and an., vol. 23, p. 26 - 86.

[4]. Protter, M.H.; Weinberger, H.F. : "Maximum principles in differential equations", 1967, Prentice Hall.

[5]. Agmon, S.; Douglis, A.; Nirenberg, N. : "Estimates near the boundary for solutions of elliptic partial differential equations satisfying general boundary conditions, I", 1959, Comm. Pure Appl. Math., vol. 12, p. 623 - 727.

[6]. Besjes, J.G. : "Singular perturbation problems for partial differential equations", thesis, 1971, Un. of Delft.

[7]. Miranda, C. : "Partial differential equations of elliptic type", 1970, Springer.

[8]. Berger, M.S.; Fraenkel, L.E. : "On the asymptotic solution of a non-linear Dirichlet problem". 1970, J. Math. Mech., vol. 19, p. 553 - 585.

[9], Berger, M.S.; Fraenkel, L.E. : "On singular perturbations of nonlinear operator equations", 1971, Indiana Un. Math. J., vol. 20, no. 7, p. 623 - 631.

AN ASYMPTOTIC THEORY FOR A CLASS OF WEAKLY NON-LINEAR OSCILLATIONS

H.W. HOOGSTRATEN

Department of Mathematics, University of Groningen,
Groningen, the Netherlands.

Abstract: For a class of weakly non-linear oscillations involving a small parameter ε we determine asymptotic solutions as $\varepsilon \downarrow 0$ which are uniformly valid on some time interval. First, we consider a general initial-value problem in $\mathbb{R}^n$ containing a small parameter ε. We derive sufficient conditions for asymptotic correctness as $\varepsilon \downarrow 0$ to be satisfied by formal asymptotic solutions. Next, we consider for the original problem formal asymptotic solutions of a two-variable type. For this type of formal asymptotic solutions the conditions for asymptotic correctness take a form which is very useful in the subsequent development of a construction technique for asymptotic solutions.

1. INTRODUCTION

Consider a class of weakly non-linear oscillations described by the following initial-value problem containing a small non-negative perturbation parameter ε:

$$w''(t,\varepsilon)+w(t,\varepsilon) = \varepsilon f\{w(t,\varepsilon),\ w'(t,\varepsilon),\varepsilon t,\varepsilon\}, \qquad t \geqq 0, \tag{1.1a}$$

$$w(0,\varepsilon) = \alpha_1(\varepsilon),\ w'(0,\varepsilon) = \alpha_2(\varepsilon). \tag{1.1b}$$

For a broad range of functions f, α_1 and α_2 the solution of (1.1) for small $\varepsilon \geqq 0$ may be expected to be a slowly varying oscillatory function of t, characterized by the presence of two time scales: a rapid scale accounting for the local periodical behaviour and a slow scale accounting for the slow modulation. It turns out to be convenient to recast the problem in terms of the slow variable $\tau = \varepsilon t$. Setting $y(\tau,\varepsilon) = w(\tau/\varepsilon,\varepsilon)$, we get

$$\varepsilon^2 y''(\tau,\varepsilon)+y(\tau,\varepsilon) = \varepsilon f\{y(\tau,\varepsilon),\ \varepsilon y'(\tau,\varepsilon),\tau,\varepsilon\}, \qquad \tau \in I, \tag{1.2a}$$

$$y(0,\varepsilon) = \alpha_1(\varepsilon),\ y'(0,\varepsilon) = \varepsilon^{-1}\alpha_2(\varepsilon), \tag{1.2b}$$

where from now on a prime indicates differentiation with respect to τ. The interval I is some τ-interval to be specified below.

We are interested in functions which approximate the solution ϕ of problem (1.2) asymptotically as $\varepsilon \downarrow 0$. A function $\tilde{\phi}_N$ will be called an Nth order asymptotic solution of (1.2) on the interval I provided ϕ exists uniquely on I for sufficiently small $\varepsilon > 0$ and positive numbers $\overline{K}$ and $\overline{\varepsilon}$ exist, independent of τ and ε, such that

$$|\phi(\tau,\varepsilon)-\tilde{\phi}_N(\tau,\varepsilon)| \leqq \overline{K}\varepsilon^{N+1},\quad |\phi'(\tau,\varepsilon)-\tilde{\phi}'_N(\tau,\varepsilon)| \leqq \overline{K}\varepsilon^{N}$$

for $\tau \in I$, $\varepsilon \in [0,\overline{\varepsilon}]$.

Under suitable differentiability conditions to be imposed on f, α_1 and α_2 we will indicate for any non-negative integer N how to determine an Nth order asymptotic solution to problem (1.2) which is uniformly valid on the τ-interval I = [0,L]. L denotes an arbitrary fixed positive number. At the same time the existence and uniqueness of the solution ϕ of (1.2) will be established on this interval I for sufficiently small $\varepsilon > 0$.

The τ-interval [0,L] corresponds to the interval $[0,L/\varepsilon]$ for the original time variable t. There are cases for which one can obtain asymptotic solutions which are actually valid on the whole infinite time interval $[0,\infty)$, for instance the weakly damped linear oscillator studied by Reiss [4]. It may be shown, however, by means of simple examples that for most problems it is impossible to obtain asymptotic solutions valid in the usual sense (that is, in the uniform norm) on the infinite interval $[0,\infty)$. Furthermore, our class of problems (1.2) contains examples for which it is impossible to obtain asymptotic solutions valid on τ-intervals larger than [0,L]. Such an example is provided by the linear oscillator with weak constant negative damping which has solutions representing slowly amplifying oscillations with amplitude proportional to e^{τ}. This shows that, in general, the interval [0,L] is the best possible τ-interval.

For the determination of asymptotic solutions of problems involving differential equations with a small parameter, one usually proceeds by first constructing so-called formal asymptotic solutions. An Nth order formal asymptotic solution ϕ_N of problem (2.1) on I is a uniformly bounded function satisfying the differential equation (1.2a) and the initial conditions (1.2b) up to a certain degree of asymptotic accuracy as $\varepsilon \downarrow 0$:

$$\varepsilon^2\phi_N''+\phi_N-\varepsilon f(\phi_N,\varepsilon\phi_N',\tau,\varepsilon) \overset{df}{=\!=} \varepsilon^{N+1}g_N(\tau,\varepsilon) = O(\varepsilon^{N+1}) \text{ uniformly for } \tau \in I,$$

$$\alpha_1(\varepsilon)-\phi_N(0,\varepsilon) = O(\varepsilon^{N+1}),\ \varepsilon^{-1}\alpha_2(\varepsilon)-\phi_N'(0,\varepsilon) = O(\varepsilon^N).$$

In Mitropol'skii's book [3] formal asymptotic solutions of a certain type are constructed for problem (1.2) and their asymptotic correctness is proved for $\tau \in I$. From a practical point of view the method of Mitropol'skii has the disadvantage that for each N the computation of an Nth order asymptotic solution involves the solution of a different non-linear first-order differential equation for the amplitude function. This equation becomes increasingly complex as N increases and this may render an analytical evaluation of asymptotic solutions impossible. The well-known two-variable construction technique for formal asymptotic solutions developed by Kevorkian and Cole (c.f. Cole [1]) does not have the disadvantage of Mitropol'skii's method. This method, however, is only applicable to weakly non-linear autonomous oscillations in general.

We will construct an Nth order formal asymptotic solution to problem (1.2) in the two-variable form

$$\phi_N(\tau,\varepsilon) = A^{(N)}(\tau,\varepsilon)\cos\{t+B^{(N)}(\tau,\varepsilon)\} + \sum_{\nu=1}^{N} \varepsilon^{\nu}U_{\nu}^{(N)}(t,\tau,\varepsilon),\quad t = \frac{\tau}{\varepsilon}, \tag{1.3a}$$

$$\text{where } A^{(N)}(\tau,\varepsilon) = \sum_{\nu=0}^{N} \varepsilon^{\nu}A_{\nu}(\tau),\ B^{(N)}(\tau,\varepsilon) = \sum_{\nu=0}^{N} \varepsilon^{\nu}B_{\nu}(\tau). \tag{1.3b}$$

Before outlining the actual determination of an Nth order formal asymptotic solution of the type (1.3), we first derive sufficient additional conditions so that it will be an Nth order asymptotic solution on the interval I. In Section 2 we prove Theorem 1 which provides such sufficient additional conditions for asymptotic correctness for a general class of initial-value problems in $\mathbb{R}^n$ under the restriction $N \geq 1$. For a subclass of weakly non-linear problems in $\mathbb{R}^n$ similar conditions, valid for $N \geq 0$, are provided by Theorem 1 bis. At the same time both theorems establish the existence and uniqueness of the exact solution on I for sufficiently small $\varepsilon > 0$. In Section 3 we consider for the original problem (1.2) a class of Nth order formal asymptotic solutions ϕ_N of the two-variable type $\phi_N(\tau,\varepsilon) = \phi_N^*(t,\tau,\varepsilon)$, $t = \tau/\varepsilon$, where ϕ_N^* is 2π-periodic in t. The sufficient additional conditions for asymptotic correctness then reduce to the conditions that for $\varepsilon \downarrow 0$

$$\int_0^{2\pi} g_N^*(t,\tau,\varepsilon)\left\{\begin{matrix}\cos t\\ \sin t\end{matrix}\right\}dt = O(\varepsilon) \text{ uniformly for } \tau \in I, \tag{1.4}$$

where $g_N^*(\tau/\varepsilon,\tau,\varepsilon) = g_N(\tau,\varepsilon)$. In Section 4 the construction of an Nth order formal asymptotic solution of the type (1.3) is developed in such a way that ϕ_N becomes of the type considered in Section 3 and so that it satisfies conditions (1.4). Then we know that it is an Nth order asymptotic solution of problem (1.2) on the interval I.

The material presented here is based on a paper by H.W. Hoogstraten and B. Kaper [2] to which the reader is referred for further details.

2. A GENERAL INITIAL-VALUE PROBLEM IN $\mathbb{R}^n$.

Consider for small $\varepsilon > 0$ the initial-value problem

$$\varepsilon x_\varepsilon'(\tau) = F_\varepsilon\{x_\varepsilon(\tau),\tau\},\ x_\varepsilon(0) = \alpha_\varepsilon,\quad \tau \in I = [0,L], \tag{2.1}$$

where $x_\varepsilon(\tau)$, $F_\varepsilon(.,\tau)$, $\alpha_\varepsilon \in \mathbb{R}^n$. The subscript ε indicates dependence on ε. We assume that for small $\varepsilon > 0$ the function $F_\varepsilon(z,\tau)$, $z \in \mathbb{R}^n$, and its first- and second-order derivatives with respect to the components of z belong to $C^0(\mathbb{R}^n\times I, \mathbb{R}^n)$.

<u>Definition 1</u>. Assume that for small $\varepsilon > 0$ problem (2.1) has a unique solution η_ε on the interval I. Let N be a nonnegative integer. An Nth order asymptotic solution of (2.1) on I is a function $\tilde{u}_\varepsilon$ satisfying

(i) $\tilde{u}_\varepsilon(\tau) \in \mathbb{R}^n$ for $\tau \in I$ and small $\varepsilon > 0$,

(ii) $\tilde{u}_\varepsilon(\tau)-\eta_\varepsilon(\tau) = O(\varepsilon^{N+1})$ uniformly for $\tau \in I$.

(When a vector function satisfies an order relation it is to be understood that the norm of the vector function satisfies the order relation. The norm of a vector or a matrix is defined as the sum of the absolute values of its elements. The order symbol O has its usual meaning and is always understood to be related to the limit process $\varepsilon \downarrow 0$).

<u>Definition 2</u>. An Nth order formal asymptotic solution of problem (2.1) on the interval I is a function u_ε satisfying

(i) $u_\varepsilon \in C^1(I, \mathbb{R}^n)$ for small $\varepsilon > 0$, $u_\varepsilon = O(1)$ uniformly for $\tau \in I$,

(ii) $\varepsilon u_\varepsilon'(\tau)-F_\varepsilon\{u_\varepsilon(\tau),\tau\} \overset{df}{=\!=} \varepsilon^{N+1}g_\varepsilon(\tau) = O(\varepsilon^{N+1})$ uniformly for $\tau \in I$, (2.2)

(iii) $\alpha_\varepsilon-u_\varepsilon(0) \overset{df}{=\!=} \varepsilon^{N+1}\tilde{\alpha}_\varepsilon = O(\varepsilon^{N+1})$. (2.3)

The following theorem provides sufficient additional conditions for an Nth order formal asymptotic solution ($N \geq 1$) to be an Nth order asymptotic solution on I. At the same time it establishes the existence and uniqueness of the solution of problem (2.1) on I for sufficiently small $\varepsilon > 0$.

Theorem 1. Let u_ε be an Nth order formal asymptotic solution ($N \geq 1$) of problem (2.1) on the interval I and let the residual function g_ε be defined by (2.2). Consider the fundamental matrix solution Ψ_ε of the linear variational equation

$$\varepsilon z_\varepsilon'(\tau) = \nabla F_\varepsilon\{u_\varepsilon(\tau),\tau\}z_\varepsilon(\tau), \quad \tau \in I, \tag{2.4}$$

satisfying $\Psi_\varepsilon(0) = E$ (unit matrix).

If positive numbers β and M exist, independent of τ and ε, such that for small $\varepsilon > 0$

$$|\Psi_\varepsilon(\tau)|,\ |\Psi_\varepsilon^{-1}(\tau)| \leq \beta, \quad \tau \in I, \tag{2.5}$$

$$\left|\int_0^\tau \Psi_\varepsilon^{-1}(s)g_\varepsilon(s)ds\right| \leq \varepsilon M,\ \tau \in I, \tag{2.6}$$

then problem (2.1) has a unique solution on I. Moreover, u_ε is an Nth order asymptotic solution of problem (2.1) on I.

Proof. The theorem will be proved by showing that for small $\varepsilon > 0$ a unique solution of (2.1) exists on I of the form $\eta_\varepsilon = u_\varepsilon + \rho_\varepsilon$, where the remainder function $\rho_\varepsilon(\tau) = O(\varepsilon^{N+1})$ uniformly for $\tau \in I$. From equations (2.1 - 2.3) it follows that ρ_ε satisfies the initial-value problem

$$\varepsilon\rho_\varepsilon'(\tau) = \nabla F_\varepsilon\{u_\varepsilon(\tau),\tau\}\rho_\varepsilon(\tau) + r_\varepsilon\{u_\varepsilon(\tau),\rho_\varepsilon(\tau),\tau\} - \varepsilon^{N+1}g_\varepsilon(\tau), \tag{2.7a}$$

$$\rho_\varepsilon(0) = \varepsilon^{N+1}\tilde{\alpha}_\varepsilon, \tag{2.7b}$$

where $\quad r_\varepsilon(u_\varepsilon,\rho_\varepsilon,\tau) = F_\varepsilon(u_\varepsilon+\rho_\varepsilon,\tau) - F_\varepsilon(u_\varepsilon,\tau) - \nabla F_\varepsilon(u_\varepsilon,\tau)\rho_\varepsilon.$

By virtue of the assumed properties of F_ε and u_ε, the variational equation (2.4) has continuous coefficients for small $\varepsilon > 0$. Then, by standard theorems for linear ordinary differential equations we know that equation (2.4) possesses for small $\varepsilon > 0$ a unique, invertable fundamental matrix solution $\Psi_\varepsilon(\tau)$ on the interval I satisfying $\Psi_\varepsilon(0) = E$.

The initial-value problem (2.7) may now be transformed uniquely into a non-linear Volterra integral equation:

$$\rho_\varepsilon(\tau) = \varepsilon^{N+1}\Psi_\varepsilon(\tau)\tilde{\alpha}_\varepsilon + \Psi_\varepsilon(\tau)\int_0^\tau \Psi_\varepsilon^{-1}(s)[\tfrac{1}{\varepsilon}r_\varepsilon\{u_\varepsilon(s),\rho_\varepsilon(s),s\} - \varepsilon^N g_\varepsilon(s)]ds. \tag{2.8}$$

This integral equation will be considered as an operator equation $\rho_\varepsilon = V\rho_\varepsilon$, where $V: C^0(I, \mathbb{R}^n) \to C^0(I, \mathbb{R}^n)$ is a non-linear Volterra integral operator. If we introduce the norm

$$\|\rho\| = \max_{\tau \in I} |\rho(\tau)|, \quad \rho \in C^0(I, \mathbb{R}^n),$$

the function space $C^0(I, \mathbb{R}^n)$ becomes a Banach space. We will apply the Banach contraction mapping principle to the ball

$$B(K) = \{\rho_\varepsilon | \rho_\varepsilon \in C^0(I, \mathbb{R}^n),\ \|\rho_\varepsilon\| \leq K\varepsilon^{N+1} \text{ for small } \varepsilon > 0\},$$

where K is a positive number (independent of ε) to be specified later.

Consider an element $\rho_\varepsilon \in B(K)$. Then, using the assumed differentiability properties of F_ε we may derive

$$|r_\varepsilon(u_\varepsilon,\rho_\varepsilon,\tau)| \leq \tilde{M}|\rho_\varepsilon|^2, \tag{2.9}$$

where $\tilde{M}$ is a positive number independent of τ and ε. By virtue of (2.3), (2.5), (2.6), (2.8) and (2.9) we have for small $\varepsilon > 0$ the estimate

$$|V\rho_\varepsilon(\tau)| \leqq \varepsilon^{N+1}\beta\alpha^* + \beta^2 L\tilde{M}K^2\varepsilon^{2N+1} + \varepsilon^{N+1}\beta M, \quad \tau \in I,$$

where the positive number α^* (independent of ε) denotes an upper bound for $|\tilde{\alpha}_\varepsilon|$. Choosing $K > \beta(\alpha^*+M)$ we have for $N \geqq 1$ and small $\varepsilon > 0$ the estimate $|V\rho_\varepsilon(\tau)| \leqq K\varepsilon^{N+1}$ uniformly for $\tau \in I$. Thus, V maps B(K) into itself.

Next, consider two elements ρ_ε and $\bar{\rho}_\varepsilon$ of B(K). Then we have

$$V\rho_\varepsilon(\tau) - V\bar{\rho}_\varepsilon(\tau) = \frac{1}{\varepsilon}\Psi_\varepsilon(\tau)\int_0^\tau \Psi_\varepsilon^{-1}(s)[r_\varepsilon\{u_\varepsilon(s),\rho_\varepsilon(s),s\} - r_\varepsilon\{u_\varepsilon(s),\bar{\rho}_\varepsilon(s),s\}]ds. \quad (2.10)$$

We make use of the following estimate:

$$|r_\varepsilon(u_\varepsilon,\rho_\varepsilon,\tau) - r_\varepsilon(u_\varepsilon,\bar{\rho}_\varepsilon,\tau)| =$$
$$= \left|\int_0^1 [\nabla F_\varepsilon\{u_\varepsilon+\xi\rho_\varepsilon+(1-\xi)\bar{\rho}_\varepsilon,\tau\} - \nabla F_\varepsilon(u_\varepsilon,\tau)][\rho_\varepsilon-\bar{\rho}_\varepsilon]d\xi\right|$$
$$\leqq \tilde{M}\{\xi_0|\rho_\varepsilon| + (1-\xi_0)|\bar{\rho}_\varepsilon|\}|\rho_\varepsilon-\bar{\rho}_\varepsilon|, \quad \xi_0 \in [0,1], \ \tau \in I, \quad (2.11)$$

where we have applied the mean-value theorem for integrals.
From (2.5), (2.10) and (2.11) we obtain

$$|V\rho_\varepsilon(\tau) - V\bar{\rho}_\varepsilon(\tau)| \leqq \beta^2 L\tilde{M}K\varepsilon^N \|\rho_\varepsilon - \bar{\rho}_\varepsilon\|, \quad \tau \in I.$$

Hence, for small $\varepsilon > 0$ and $N \geqq 1$, a positive number $\lambda < 1$ exists independent of ε such that for any ρ_ε and $\bar{\rho}_\varepsilon$ belonging to B(K):

$$\|V\rho_\varepsilon - V\bar{\rho}_\varepsilon\| \leqq \lambda\|\rho_\varepsilon - \bar{\rho}_\varepsilon\|,$$

that is, V is a contraction mapping of B(K) into itself. Then, by virtue of the Banach contraction mapping principle a unique fixed point $\tilde{\rho}_\varepsilon \in B(K)$ of V exists. This completes the proof of Theorem 1.

For a class of weakly non-linear problems we have the following theorem in which the restriction $N \geqq 1$ has been removed.

Theorem 1 bis. Let u_ε be an Nth order formal asymptotic solution of the initial-value problem

$$\varepsilon x_\varepsilon'(\tau) = \tilde{A}_\varepsilon(\tau)x_\varepsilon(\tau) + \varepsilon f_\varepsilon\{x_\varepsilon(\tau),\tau\}, \quad x_\varepsilon(0) = \alpha_\varepsilon, \quad \tau \in I, \quad (2.12)$$

where $\tilde{A}_\varepsilon$ is an nxn matrix of elements belonging to $C^0(I, \mathbb{R})$ for small $\varepsilon > 0$ and where f_ε has the same differentiability properties as those assumed for F_ε at the beginning of this section. Let g_ε be the residual function corresponding to u_ε and let Ψ_ε denote the fundamental matrix solution of the linear variational equation

$$\varepsilon z_\varepsilon'(\tau) = \tilde{A}_\varepsilon(\tau)z_\varepsilon(\tau) + \varepsilon\nabla f_\varepsilon\{u_\varepsilon(\tau),\tau\}z_\varepsilon(\tau), \quad \tau \in I, \quad (2.13)$$

satisfying $\Psi_\varepsilon(0) = E$.

If positive numbers β and M exist, independent of τ and ε, such that the estimates (2.5) and (2.6) hold for small $\varepsilon > 0$, then problem (2.12) has a unique solution on the interval I. Moreover, u_ε is an Nth order asymptotic solution of (2.12) on I.

Proof. The proof is nearly identical with the proof of Theorem 1. The crucial difference is that the coefficient of r_ε in square brackets in equation (2.8) is now equal to 1 instead of ε^{-1}. This renders the restriction $N \geqq 1$ superfluous in the remaining part of the proof.

3. FORMAL ASYMPTOTIC SOLUTIONS OF THE TWO-VARIABLE TYPE.

In this section we return to the original weakly non-linear initial-value problem (1.2) which can be written in the vector form (2.12) where

$$x_\varepsilon(\tau) = \begin{pmatrix} y(\tau,\varepsilon) \\ \varepsilon y'(\tau,\varepsilon) \end{pmatrix}, \quad \tilde{A} = \begin{pmatrix} 0 & 1 \\ -1 & 0 \end{pmatrix}, \quad \alpha_\varepsilon = \begin{pmatrix} \alpha_1(\varepsilon) \\ \alpha_2(\varepsilon) \end{pmatrix},$$

$$f_\varepsilon\{x_\varepsilon(\tau),\tau\} = \begin{pmatrix} 0 \\ f\{y(\tau,\varepsilon),\varepsilon y'(\tau,\varepsilon),\tau,\varepsilon\} \end{pmatrix}.$$

A function ϕ_N will be called an Nth order formal asymptotic solution of problem (1.2) on I if the corresponding vector function $u_\varepsilon(\tau) = \mathrm{col}\{\phi_N(\tau,\varepsilon), \varepsilon\phi_N'(\tau,\varepsilon)\}$ is an Nth order formal asymptotic solution of problem (2.12) in the sense of Definition 2. Similarly we define an Nth order asymptotic solution of (1.2) on I.

From now on we assume that α_1 and α_2 are infinitely differentiable functions of ε for small $\varepsilon \geqq 0$ and that f is an infinitely differentiable function of its four arguments in the region $\mathbb{R}^2\times I\times[0,\varepsilon_0]$. Here, ε_0 denotes a generic positive number, that is, ε_0 will not necessarily be the same number each time it appears.

Consider a formal asymptotic solution ϕ_N of the two-variable type, that is,

$$\phi_N(\tau,\varepsilon) = \phi_N^*(t,\tau,\varepsilon), \quad t = \tau/\varepsilon,$$

where (i) $\phi_N^* \in C^\infty(\mathbb{R}\times I\times[0,\varepsilon_0], \mathbb{R})$,

(ii) $\phi_N^*(t+2\pi,\tau,\varepsilon) = \phi_N^*(t,\tau,\varepsilon)$ for $(t,\tau,\varepsilon) \in \mathbb{R}\times I\times[0,\varepsilon_0]$.

(A function of t, τ and ε satisfying (i) and (ii) will be said to belong to the class P^∞).

If we consider the above two-variable type of formal asymptotic solutions for problem (1.2) we know (c.f. Lemma 1 in Hoogstraten and Kaper [2]) that the fundamental matrix solution Ψ_ε of the corresponding variational equation in $\mathbb{R}^2$ consists of elements of the form

$$\tilde{C}(\tau,\varepsilon)\cos\frac{\tau}{\varepsilon} + \tilde{D}(\tau,\varepsilon)\sin\frac{\tau}{\varepsilon} + O(\varepsilon) \text{ uniformly for } \tau \in I,$$

where the coefficient functions $\tilde{C}$ and $\tilde{D}$ belong to the class $C^\infty(I\times[0,\varepsilon_0], \mathbb{R})$.

Combining Theorem 1 bis and the above-mentioned Lemma 1, we may prove the following main theorem ([2]).

Theorem 2. Consider the initial-value problem (1.2) where α_1 and α_2 are infinitely differentiable functions of ε for small $\varepsilon \geqq 0$ and where f is an infinitely differentiable function of its four arguments in the region $\mathbb{R}^2\times I\times[0,\varepsilon_0]$. Assume that ϕ_N is an Nth order formal asymptotic solution of problem (1.2) on the interval I of the two-variable type

$$\phi_N(\tau,\varepsilon) = \phi_N^*(t,\tau,\varepsilon), \quad t = \tau/\varepsilon, \quad \phi_N^* \in P^\infty.$$

Then one can write

$$\varepsilon^2\phi_N''(\tau,\varepsilon)+\phi_N(\tau,\varepsilon)-\varepsilon f\{\phi_N(\tau,\varepsilon),\varepsilon\phi_N'(\tau,\varepsilon),\tau,\varepsilon\} = \varepsilon^{N+1}g_N^*(t,\tau,\varepsilon)$$

with $g_N^* \in P^\infty$. Moreover, if

$$\int_0^{2\pi} g_N^*(t,\tau,\varepsilon)\left\{\begin{matrix}\sin t\\ \cos t\end{matrix}\right\}dt = O(\varepsilon) \text{ uniformly for } \tau \in I, \tag{3.1}$$

then problem (1.2) has a unique solution on I and ϕ_N is an Nth order asymptotic solution on I.

4. A CONSTRUCTION METHOD FOR ASYMPTOTIC SOLUTIONS.

We will outline a two-variable construction technique leading to an Nth order formal asymptotic solution $\phi_N(\tau,\varepsilon)$ of problem (1.2) on the interval I satisfying the conditions of Theorem 2. The function ϕ_N will be constructed in the form of a generalized finite asymptotic perturbation series involving functions of t, τ and ε:

$$\phi_N(\tau,\varepsilon) = \sum_{\nu=0}^{N} \varepsilon^\nu U_\nu^{(N)}(t,\tau,\varepsilon). \tag{4.1}$$

For any integer $N \geq 0$ and $\nu = 0,1,2,\ldots,N$ we assume that $U_\nu^{(N)}$ belongs to $C^\infty(\mathbb{R}\times I\times[0,\varepsilon_0], \mathbb{R})$ and is uniformly bounded as $\varepsilon \downarrow 0$ together with its t- and τ-derivatives up to any order. The residual expression

$$\varepsilon^2\phi_N''+\phi_N-\varepsilon f(\phi_N,\varepsilon\phi_N',\tau,\varepsilon) = U_{0tt}^{(N)}+U_0^{(N)}+\varepsilon(\ldots)+\ldots\ , \quad t = \tau/\varepsilon, \tag{4.2}$$

should be $O(\varepsilon^{N+1})$ uniformly for $\tau \in I$. As a first step we put

$$U_{0tt}^{(N)}+U_0^{(N)} = 0.$$

The general solution of this equation for $U_0^{(N)}$ is taken in the form

$$U_0^{(N)}(t,\tau,\varepsilon) = A^{(N)}(\tau,\varepsilon)\cos\{t+B^{(N)}(\tau,\varepsilon)\}, \tag{4.3}$$

where the unknown functions $A^{(N)}$ and $B^{(N)}$ should belong to $C^\infty(I\times[0,\varepsilon_0], \mathbb{R})$. Note, then, that $U_0^{(N)} \in P^\infty$. We construct $A^{(N)}$ and $B^{(N)}$ of the form

$$A^{(N)}(\tau,\varepsilon) = \sum_{\nu=0}^{N} \varepsilon^\nu A_\nu(\tau), \quad B^{(N)}(\tau,\varepsilon) = \sum_{\nu=0}^{N} \varepsilon^\nu B_\nu(\tau), \tag{4.4}$$

with A_ν, $B_\nu \in C^\infty(I, \mathbb{R})$. So we have

$$\phi_N(\tau,\varepsilon) = A^{(N)}(\tau,\varepsilon)\cos\{t+B^{(N)}(\tau,\varepsilon)\} + \sum_{\nu=1}^{N} \varepsilon^\nu U_\nu^{(N)}(t,\tau,\varepsilon), \tag{4.5}$$

where $A^{(N)}$ and $B^{(N)}$ are given by (4.4). The determination of the functions $U_\nu^{(N)}$, A_ν and B_ν is illustrated by means of the example

$$\varepsilon^2 y''(\tau,\varepsilon)+y(\tau,\varepsilon)-\varepsilon a(\tau)y^2(\tau,\varepsilon)-\varepsilon^2 b(\tau)y'(\tau,\varepsilon) = 0, \quad \tau \in I, \tag{4.6a}$$

$$y(0,\varepsilon) = \alpha, \quad y'(0,\varepsilon) = 0, \tag{4.6b}$$

where a, b $\in C^\infty(I, \mathbb{R})$. We now substitute for y the expansion (4.5) into the lefthand side of (4.6a) and expand the result in powers of ε, leaving $\sin\{t+B^{(N)}\}$, $\cos\{t+B^{(N)}\}$ and higher harmonics intact. Upon equating to zero the contributions of order $O(\varepsilon)$, $O(\varepsilon^2),\ldots,O(\varepsilon^N)$ we get a set of recursive differential equations for the functions $U_1^{(N)}$, $U_2^{(N)},\ldots U_N^{(N)}$:

$$U_{1tt}+U_1 = (2A_0'-bA_0)\sin(t+B)+2A_0B_0'\cos(t+B)+\tfrac{1}{2}aA_0^2\{1+\cos 2(t+B)\}, \tag{4.7}$$

$$\begin{aligned} U_{2tt}+U_2 &= (2A_1'+2A_0'B_0'+A_0B_0'')\sin(t+B)+\{2A_0B_1'+2A_1B_0'-A_0''+A_0(B_0')^2\}\cos(t+B) \\ &+ 2aA_0\{U_1+A_1\cos(t+B)\}\cos(t+B)+b\{U_{1t}+A_0'\cos(t+B)-(A_0B_0'+A_1)\sin(t+B)\}, \end{aligned} \tag{4.8}$$

$$U_{3tt}+U_3 = \ldots, \text{ etc.},$$

where, for convenience, the superscript (N) has been omitted.

The boundedness condition on U_1 leads to the suppression of the "secular" terms involving $\sin(t+B)$ and $\cos(t+B)$ in the righthand side of (4.7). This yields equations for A_0 and B_0:

$$2A_0'(\tau)-b(\tau)A_0(\tau) = 0, \quad B_0'(\tau) = 0. \tag{4.9}$$

Solving (4.7) we get for U_1:

$$U_1(t,\tau,\varepsilon) = \tfrac{1}{2}a(\tau)A_0^2(\tau)[1-\tfrac{1}{3}\cos 2\{t+B(\tau,\varepsilon)\}]\in P^\infty. \tag{4.10}$$

Note that no solutions sin t and cos t of the homogeneous equation corresponding to (4.7) have been included in (4.10). This is possible because of the freedom introduced in (4.3) by the expansions (4.4).

Similarly we determine recursively the functions $U_2,\ldots,U_N \in P^\infty$. The suppression of secular terms in the equations for $U_2,\ldots,U_N$ leads to equations for A_1, B_1, A_2, $B_2,\ldots,A_{N-1}$, B_{N-1}. In this way we achieve that (4.1) can be written as a function $\phi_N^*(t,\tau,\varepsilon)$ belonging to P^∞. Furthermore, the corresponding residual expression (4.2) is of the form

$$\varepsilon^{N+1}G_N(t,\tau,\varepsilon)+O(\varepsilon^{N+2}), \quad G_N \in P^\infty,$$

and hence it is $O(\varepsilon^{N+1})$ uniformly for $\tau \in I$. It is not difficult to see that if we would determine a $(N+1)^{th}$ order formal asymptotic solution, the function G_N would figure as the righthand side of the equation for U_{N+1}. The function G_N still contains A_N and B_N as unknowns. The additional conditions (3.1) of Theorem 2 are satisfied by the suppression of the "secular" terms in G_N. This yields equations for A_N and B_N. The initial conditions for the equations for A_0, $B_0,\ldots,A_N$, B_N are obtained by expanding the expressions $\phi_N(0,\varepsilon)-\alpha$ and $\varepsilon\phi_N'(0,\varepsilon)$ in powers of ε and equating to zero the contributions of order $O(1)$, $O(\varepsilon),\ldots,O(\varepsilon^N)$.

To illustrate this, we determine a first-order asymptotic solution for the initial-value problem (4.6). In order to satisfy conditions (3.1) of Theorem 2 we substitute (4.10) into the righthand side of the equation (4.8) for U_2. The suppression of the secular terms yields the equations

$$2A_1'(\tau)-b(\tau)A_1(\tau) = 0, \; 2A_0(\tau)B_1'(\tau)-A_0''(\tau)+b(\tau)A_0'(\tau)+\tfrac{5}{6}a^2(\tau)A_0^3(\tau) = 0. \tag{4.11}$$

Thus, a first-order asymptotic solution ϕ_1 of (4.6) on I is given by

$$\begin{aligned} \phi_1(\tau,\varepsilon) &= \{A_0(\tau)+\varepsilon A_1(\tau)\}\cos\{\tfrac{\tau}{\varepsilon}+B_0(\tau)+\varepsilon B_1(\tau)\} + \\ &+ \tfrac{1}{2}\varepsilon a(\tau)A_0^2(\tau)[1-\tfrac{1}{3}\cos 2\{\tfrac{\tau}{\varepsilon}+B_0(\tau)+\varepsilon B_1(\tau)\}], \end{aligned} \tag{4.12}$$

with A_0, A_1, B_0, B_1 solutions of equations (4.9) and (4.11). The initial conditions for equations (4.9) and (4.11) are obtained by expanding the expressions $\phi_1(0,\varepsilon)-\alpha$ and $\varepsilon\phi_1'(0,\varepsilon)$ in powers of ε and equating to zero the contributions of order $O(1)$ and $O(\varepsilon)$:

$$A_0(0) = \alpha,\ B_0(0) = 0,\ A_1(0) = -\frac{1}{3}\alpha^2 a(0),\ B_1(0) = \alpha^{-1}A_0'(0). \qquad (4.13)$$

The solutions of equations (4.9) and (4.11) satisfying (4.13) are

$$A_0(\tau) = \alpha \exp\left\{\frac{1}{2}\int_0^\tau b(s)ds\right\},\quad B_0(\tau) = 0,$$

$$A_1(\tau) = -\frac{1}{3}\alpha^2 a(0) \exp\left\{\frac{1}{2}\int_0^\tau b(s)ds\right\},$$

$$B_1(\tau) = \frac{1}{4}b(0)+\frac{1}{4}b(\tau) - \int_0^\tau \left\{\frac{1}{8}b^2(s)+\frac{5}{12}a^2(s)A_0^2(s)\right\} ds.$$

REFERENCES.

1. J.D. Cole, Perturbation methods in applied mathematics, Ginn/Blaisdell 1968.
2. H.W. Hoogstraten and B. Kaper, An asymptotic theory for a class of weakly non-linear oscillations, to appear in the Arch. for Rat. Mech. and Anal. (1974).
3. Yu. A. Mitropol'skii, Problems of the asymptotic theory of non-stationary vibrations, Israel Program for Scientific Translations, Jerusalem 1965.
4. E.L. Reiss, On multivariable asymptotic expansions, SIAM Review 13 (1971) 189.

Adams, R.A., 118
Agmon, S., 195
Akhiezer, N.I., 28
Arscott, F.M., 94, 102

Bennewitz, C., 12
Berger, J.G., 195
Besjes, J.G., 195
Böcker, U., 127
Boer, W.L., 109
Bogoliubov, N.N., 162
Bremerman, H., 147
Brinck, I., 78
Browder, F.E., 147
Browne, P.J., 94

Carmichael, R., 147
Chikwendu, S.C., 162
Clark, C.W., 118
Coddington, E.A., 78, 162
Cole, J.D., 205
Coppel, W.A., 78
Cordes, H.O., 94

Douglis, A., 195
Dyke, M.D. van, 173

Eastham, M.S.P., 66, 78, 118
Eckhardt, K.J., 127
Eckhaus, W., 162, 173, 178, 195
Edmunds, D.E., 118
Emanuelsson, K., 12
Evans, W.D., 94, 118
Everitt, W.N., 28, 66, 78

Fraenkel, L.E. , 173, 195

Garabedian, P. , 102
Giertz, M., 66
Glazman, I.M., 28, 94, 118
Grasman, J., 179
Gregus, M., 94, 102

Hartman, P., 66, 79
Hille, E., 79
Hoogstraten, H.W., 205

Ikebe, T., 66, 127
Ince, E.L., 94
Iohvidov, I.S., 28
Ismagilov, R.S., 79

Jager, E.M. de, 178, 195
Jorgens, K., 118, 127

Kaper, B., 205
Kaplun, S., 173
Kato, T., 66, 118, 127
Kevorkian, J., 162
Kimura, T., 12
Knowles, I., 79
Koppenfels, W., 102 ,
Krein, M.G., 28
Kuroda, S.T., 127, 128

Lagerstrom, P.A., 173
Levinson, N., 78, 79, 162
Lions, J.L., 147
Lyusternik, L.A., 179

Malurkar, S.L., 102
Martini, R., 109
Mc Lachlan, E.K., 102
Mc Leod, J.B., 66, 67
Miranda, C., 195
Mitropolski, I.A., 162, 205
Möglich, F., 102

Naimark, M.A., 66, 79
Neuman, F., 94, 102
Niessen, H.D., 28, 43, 56
Nirenberg, N., 195

Oehme, R., 147
Ong, K.S., 28

Pleijel, A., 12, 28
Protter, M.H., 195

Reiss, E.L., 205
Roseau, M., 162

Schäfke, F.W., 43
Schneider, A., 28, 43, 56
Schwartz, L., 147
Sears, D.B.
Shotwell, D.A., 28, 43, 56
Sips, R., 102
Sleeman, D.B., 94, 102

Takahasi, M., 12
Taylor, J.G., 147
Tayoshi, T. 127
Thompson, M.L., 78
Tillman, H.G., 147
Titchmarsh, E.C., 66, 79
Trèves, F., 147

Veselić, K., 128
Visik, M.I., 179
Vladimirov, V.S., 147
Volosov, V.M., 162

Walter, J., 79
Watson, N.G., 102
Weidmann, J., 127, 128
Weinberger, H.F., 195
Weyl, H., 12, 28, 43, 56, 66, 79
Whittaker, E.T., 102
Wintner, A., 79
Wong, J.S.W., 79

Yosida, K., 109, 128

List of Participants

France

1. Prof. Dr. J. Mauss . Université de Toulouse

Germany

2. Prof. Dr. R. Mennicken . Universität Regensburg
3. Prof. Dr. H. D. Niessen . Gesamthochschule Essen
4. Prof. Dr. A. Schneider . Gesamthochschule Wuppertal
5. Prof. Dr. F. Stummel . Johann Wolfgang Goethe-Universitat, Frankfurt am Main
6. Prof. Dr. J. Weidmann . Johann Wolfgang Goethe-Universitat, Frankfurt am Main

Sweden

7. Dr. M. Giertz . Kungl. Tekniska Högskolan, Stockholm
8. Prof. Dr. Å. Pleyel . Uppsala Universitet, Uppsala

United Kingdom

9. Prof. Dr. F. M. Arscott . University of Reading
10. Dr. M. S. P. Eastham . University of London
11. Dr. W. D. Evans . University of Cardiff
12. Prof. Dr. W. N. Everitt . University of Dundee
13. Dr. J. B. Macleod . University of Oxford
14. Dr. I. M. Michael . University of Dundee
15. Dr. B. D. Sleeman . University of Dundee

The Netherlands

1. Prof. Dr. B. L. J. Braaksma . Rijksuniversiteit Groningen
2. Dr. D. W. Bresters . Universiteit van Amsterdam
3. Prof. Dr. Ir W. Eckhaus . Rijksuniversiteit Utrecht
4. Drs. J. A. van Gelderen . Technische Hogeschool Delft

5. Dr. Ir J. de Graaf	.	Rijksuniversiteit Groningen
6. Dr. Ir J. Grasman	.	Mathematisch Centrum Amsterdam
7. Drs P.P.N. de Groen	.	Vrije Universiteit Amsterdam
8. Ir E.W.C. van Groesen	.	Universiteit van Amsterdam
9. Drs A. van Harten	.	Rijksuniversiteit Utrecht
10. Dr. Ir A. J. Hermans	.	Technische Hoeschool Delft
11. Dr. Ir H. W. Hoogstraten	.	Rijksuniversiteit Groningen
12. Prof. Dr. E. M. de Jager	.	Universiteit van Amsterdam
13. Drs B. Kaper	.	Rijksuniversiteit Groningen
14. Prof. Dr H. A. Lauwerier	.	Universiteit van Amsterdam
15. Prof. Dr. C. G. Lekkerkerker	.	Universiteit van Amsterdam
16. Dr. Ir H. Lemei	.	Technische Hogeschool Delft
17. Prof. Dr A. H. M. Levelt	.	Katholieke Universiteit Nijmegen
18. Ir M. Martini	.	Technische Hogeschool Delft
19. Prof. Dr Ir G. Y. Nieuwland	.	Vrije Universiteit Amsterdam
20. Drs H. G. J. Pijls	.	Universiteit van Amsterdam
21. Drs J. W. de Roever	.	Mathematisch Centrum Amsterdam
22. Drs G. H. Schmidt	.	Rijksuniversiteit Groningen
23. Dr. Ir H. S. V. Snoo	.	Rijksuniversiteit Groningen
24. Drs. N. M. Temme	.	Mathematisch Centrum Amsterdam
25. Dr. F. Verhulst	.	Rijksuniversiteit Utrecht